The Amateur Astronomer

Sir Patrick Moore

The Amateur Astronomer

Twelfth Edition

With 140 Figures

 Springer

Cover illustration: The Great Nebula in Orion, and lunar eclipse (inset); courtesy of David Hanon, Ringgold, GA.

British Library Cataloguing in Publication Data
Moore, Patrick
    The amateur astronomer.–12th ed.
    1. Astronomy–Popular works 2. Astronomy–Observers'
    manuals
    I. Title
    522

ISBN 13: 978-1-84996-941-3   e-ISBN 13: 978-1-84628-286-7
Springer Science+Business Media
springeronline.com

© Springer-Verlag London Limited 2010

Printed in Singapore
58/3830-543210 Printed on acid-free paper

Publisher's Note

The 16 star maps shown in Appendix 25 are the original line drawings that Patrick Moore feels are the most suitable as a guide to finding the constellations.

The star-map drawings have in the past proved a much easier guide than any printed version of them. In this new edition we start with line drawings of famous constellations: Ursa Major and Orion (Maps I and II), before moving on to the more difficult ones later (Maps IV–XVI). Indeed, the detailed maps have been of great service to the more experienced astronomy enthusiasts as well as to the newcomer.

We decided that the presentation of the maps was also important. The sizes of the stars are now smaller than in previously published diagrams. This is to allow for more space, particularly in the more detailed diagrams. For somebody starting out, there's nothing worse than having to find your way through horribly cluttered star maps! So we included the original line drawings with the aim of solving this problem.

Preface to the First Edition

Many popular books upon astronomy have been written during the past few years, but most of them cater either for the casual dabbler who is content to learn from the depths of his armchair or else for the serious amateur who already knows the main facts. What I have done, or tried to do, is to strike a happy mean. This book has been aimed at the needs of the beginner who knows nothing whatsoever, but who is nevertheless anxious to make a start with what equipment he can collect at limited cost.

All astronomers, professional or amateur, were beginners once, and all have had to draw upon the experience of those who have learned before them. I feel some diffidence about offering myself as a guide, but at least I have one qualification: in my early days as an observer I made almost every mistake that it is possible to make! This explains the frequent occurrence of such phrases as "I once saw...." and "I remember that when I...." I hope therefore that what I have written may prevent others from falling into the same ridiculous traps.

A common fault in popular books is that too much space is devoted to the Moon and planets, and too little to the greater problems of the stars. I am well aware that I have laid myself open to precisely this criticism, but there is a reason for it. I repeat that I am writing for the amateur who wants to observe; and while the owner of a small telescope can make himself extremely useful in the lunar and planetary field, he is rather more limited with regard to stellar problems. I hope, therefore, that the fault may be forgiven.

If this book has a use, it will be to the man who works with cheap and limited equipment. I have, however, given a list of more advanced works which can be consulted by anyone who wants to go more deeply into the subject.

Astronomy is the most satisfying of all hobbies; taken as a class, astronomers are friendly folk. If my book persuades a few people to take a real interest in the heavens, I shall feel that it has been well worth writing.

Patrick Moore
August 1957

Preface to the Twelfth Edition

The first edition of *The Amateur Astronomer* was published almost half a century ago. Other editions followed, and I hope it is fair to say that they introduced quite a number of people to astronomy. But things have changed since then. In 1957, the average amateur astronomer worked with a modest telescope and a simple camera; his main targets were the planets and some variable stars. Not so today. Electronic devices have largely replaced photographic film, telescopes are computer-controlled and the well-equipped camera can produce results equal to those of major professional observatories a few decades ago. Old-fashioned visual observers are as outdated as dinosaurs.

I have to admit that I am a dinosaur, and when I was asked to prepare a new edition of this book, I had to make a decision. There was no point in catering for the electronics expert and computer user; others can do that far better than I ever could. So it was better to retain the original pattern, bringing it up to date but not attempting to go further. If you belong to the technology of the twenty-first century, this is not the book for you; otherwise – well, I hope that you will find it useful.

Sir Patrick Moore
Selsey, January 2005

Contents

APPENDICES

Chapter 1

Astronomy as a Hobby

The twenty-first century is well into the Age of Science. Over the past hundred years our whole way of life has changed beyond all recognition, and if we could go back to Victorian times it would seem as strange as visiting another planet. I can well remember the time when there were no computers and no television; I can recall buying my first "wireless" and trying to tune in to the Test Matches being played in Australia. Space-travel, of course, was pure science fiction, and even flying from Britain to America was very much of an adventure.

One result of this amazing progress is that science has become specialized. It used to be possible for the amateur to make useful discoveries, while to the normal research worker the possibilities were endless; there was always something new "just around the corner". By now the amateur is much more limited. Most modern research needs equipment far too expensive to be assembled by a private individual, and even theoretical work is beyond most people who have not had detailed technical training.

Astronomy is the one science in which these limitations are not so crippling. The chances of making an important discovery are less than they used to be, but they still exist; for instance Tom Bopp, co-discoverer of the brilliant comet which graced our skies a few years ago, is purely an amateur, and so is Rev. Robert Evans, the Australian clergyman who has discovered a remarkable number of exploding stars in external galaxies. There is plenty of scope.

It is obvious that there are some branches of astronomy which cannot be tackled by the amateur, but others can, mainly because professional astronomers have neither the time or the inclination to undertake some types of routine work. To drive home this point, it may be useful to give a definite instance of what I mean, though admittedly it does go back a good many more years. In 1955 it was found that the giant planet Jupiter is a source of radio waves, and researchers were very anxious to know whether these radiations came from the planet as a whole, or from discrete surface features such as the famous Great Red Spot. They therefore appealed to the Jupiter Section of the British Astronomical Association, whose members had been making observations of the surface details and knew them extremely well. The BAA amateurs suddenly found that their patient labours over the years had become of real importance, and the question was answered: discrete

1

features are not involved. I agree that the situation today is different, but opportunities are still there. Remember, though, that sporadic and haphazard observations are of no practical value, enjoyable though they may be. One has to be methodical, and normally the observer will concentrate upon one particular field of research. In my case it happened to be the Moon; two close friends of mine hate the Moon because when near full it makes the sky so bright that dim objects such as comets are drowned.* Others concentrate upon monitoring variable stars, hunting for supernovae in outer galaxies, or taking pictures of nebulae. It all depends upon what attracts you most.

One question often asked is: "What is the real use of astronomy – and why spend time watching stars and planets when there is so much to be done on our own world?" On the face of it, the question seems rather reasonable enough, and it is not immediately obvious why the astronomer should become excited about the appearance of a white spot on Saturn, or the flaring-up of a new star in Cygnus. But astronomy is linked with all other sciences, and you cannot separate it from, say, physics or chemistry any more than you can separate arithmetic from algebra. The same argument applies to space research; recently I was visiting a hospital in Bristol, where they were scanning an unborn baby for possible defects – using equipment developed for use in space. I am always tempted to answer that question with another: "What is the practical use of a Rembrandt painting or a Beethoven symphony?" The only answer is that a great picture or a great piece of music can give enjoyment to millions of people.

So far as astronomy is concerned, it can take up as much or as little time as you like. You cannot draw the best out of it unless you are prepared to take a little trouble, and I well remember how I went about it, admittedly at the tender age of seven. I obtained an elementary book, and grasped the main facts, then I went out on every clear night and learned my way around the sky; next, I borrowed a pair of binoculars and began searching for objects such as star-clusters. The procedure worked well for me, and my enthusiasm has never waned. This book is an attempt to answer a second question which has been put to me on countless occasions: "If I want to make astronomy my hobby, how do I go about it?"

---

*During my Moon-mapping a work in the 1950s, I used to ring up the Paris observatory, where there is a 33-inch refracting telescope – one of the best in the world. "Any chance of a few nights with the 33-inch?" "Well, you can have it for a couple of nights near full moon." "Many thanks. That's when I want it." They became very used to me!

Chapter 2

The Unfolding Universe

A subject can always be better understood if something is known about its history. Though we no longer worship our "honourable ancestors", it is a distinct help to look back through time in order to see how knowledge has been built up through the centuries. This is particularly true with astronomy, which is the oldest science in the world – so old, indeed, that we do not know when it began.

Most people of today have at least some knowledge of the universe in which we live. The Earth is a globe nearly 8000 miles in diameter, and is one of nine planets revolving round the Sun. The best way of summing up the difference between a planet and a star is to say that the Earth is a typical planet, while the Sun is a typical star.

Five planets – Mercury, Venus, Mars, Jupiter and Saturn – were known to the ancients, while three more have been discovered in modern times. Jupiter is the largest of them, and its vast globe could swallow up more than a thousand bodies the volume of the Earth, but even Jupiter is tiny compared with the Sun. The stars of the night sky are themselves suns, many of them far larger and more brilliant than our own, and appearing small and faint only because they are so far away. On the other hand, the Moon shines more brilliantly than any other object in the sky apart from the Sun. Appearances are deceptive; the Moon is a very junior member of the Solar System, and it has no light of its own. It has a diameter only about one-quarter that of the Earth, and it is much the closest natural body in the sky.

The whole celestial vault seems to revolve round the Earth once in 24 hours. This apparent motion is due, of course, to the fact that the Earth is spinning on its axis from west to east. Of all the celestial objects, only the Moon genuinely moves round the Earth. We are used to taking these facts for granted, but in early times it was (rather naturally) believed that the Earth was flat and stationary. The Sun and Moon were worshipped as gods, and the appearance of anything unusual, such as a comet, was taken to be a sign of divine displeasure.

It is usually said that the first astronomers were the Chaldaeans, the Egyptians and the Chinese. In a way this is true enough; these ancient civilizations made useful records, but they had no real understanding of the nature of the universe or even of the Earth itself.

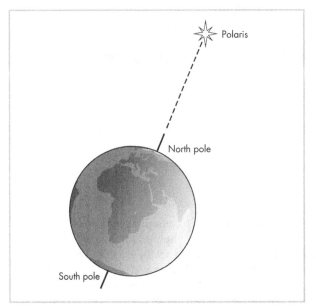

Fig. 2.1. Inclination of the Earth's axis.

The main story begins around 3000BC, when the 365-day year was first adopted in Egypt and in China. This, too, was the approximate date of the building of that remarkable structure which we know as the Great Pyramid of Cheops, still one of the world's main tourist attractions. Astronomically, it is of special interest because its main passage is aligned with what was then the north pole of the sky.

The Earth's axis of rotation is inclined at an angle of $23\frac{1}{2}$ degrees to the perpendicular to its orbit, and points northward to the celestial pole (Fig 2.1). Today the pole is marked approximately by a bright star known as Polaris, familiar to every navigator because it seems to remain almost stationary while the entire sky revolves round it. In Cheops' time, however, the polar point was in a different position, close to a much fainter star, Thuban in the constellation of the Dragon. The reason for this change is that the Earth is 'wobbling' slightly, in the manner of a gyroscope which is running down, so that the direction of the axis is describing a small circle in the sky. The effect is very slight, but the shift of the pole has become appreciable since the Pyramid was built.

The Egyptians divided up the stars into constellations, though their scheme was different from that which we follow today. They also made good measurements, and they regulated their calender by the 'heliacal rising' of the star Sirius – that is to say, the date when Sirius could first be seen in the dawn sky. Some of their other ideas were very wide of the mark. They believed the world to be rectangular, with Egypt in the middle, and that the sky was formed by the body of a goddess with the rather appropriate name of Nut.

The Chinese were equally good observers, and made careful records of comets and eclipses. Total eclipses of the Sun are particularly spectacular, and at this point I cannot resist re-telling a famous legend, even though experts assure me that it is certainly untrue! Here, then, is the story of Hsi and Ho:

The Moon revolves round the Earth once a month, while the Earth takes a year to complete one journey round the Sun. The Moon is much smaller than the Sun, but it is also much closer, so that – by pure chance – the two look almost exactly the same size. When the Sun, Moon and Earth move into an exact line, with the Moon in the mid-position, the result is a total solar eclipse. The Moon blots out the bright disk of the Sun, and for a few moments – never as long as eight minutes – we can see the glorious pearly corona and the 'red flames' or prominences; the sky becomes so dark that stars can be seen.

The Chinese knew how to predict eclipses – more or less – but they did not know that the Moon was involved; they thought that the Sun was in danger of being eaten by a hungry dragon, so that the only course was to scare the beast away by shouting, screaming, wailing, and beating gongs and drums. (It always worked!) The legend says that in 2136 BC, during the reign of the Emperor Chung K'ang, the Court Astronomers, Hsi and Ho, failed to give due warning that an eclipse was due, so that no preparations were made – and since Hsi and Ho had imperilled the whole world by their neglect of duty, they were summarily executed. I am sorry that the experts have demolished this tale. Had it been true, Hsi and Ho would have been the first known scientific martyrs in history. I have no idea where the story originated.

Astronomy in its true form began with the Greeks, who not only made observations but who also tried to explain them. The first of the great philosophers was Thales of Miletus, who was born in 624 BC; the last was Ptolemy of Alexandria, and with his death, in or about A.D. 180, the classical period of science comes to an end. During the intervening eight centuries, human thought made remarkable progress.

Thales himself may have been the first to realize that the Earth is a globe, but unfortunately all his original writings have been lost. The first definite arguments against the old idea of a flat Earth were given by Aristotle, who was born in 384 B.C.

Fig. 2.2. Eratosthenes' method of measuring the circumference of the Earth.

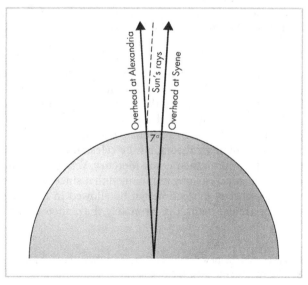

and died in 322. Aristotle was one of the most brilliant men of the ancient world, and his reasoning shows the Greek mind at its best.

As Aristotle points out, the stars appear to alter in altitude above the horizon according to the latitude of the observer. Polaris appears to remain fairly high in the sky as seen from Greece, because Greece is well north of the terrestrial equator; from Egypt, Polaris is lower; from southern latitudes it cannot be seen at all, since it never rises above the horizon. On the other hand Canopus, a brilliant star in the southern part of the sky, can be seen from Egypt but not from Greece. This is just what would be expected on the theory of a round Earth, but such behaviour cannot possibly be explained if we suppose the Earth to be flat. Aristotle also noticed that during a lunar eclipse, when the Earth's shadow falls across the Moon, the edge of the shadow appears curved – indicating that the surface of the Earth must also be curved.

The next step was taken by Eratosthenes of Cyrene, who succeeded in measuring the length of the Earth's circumference (c. 240 BC). His method was most ingenious, and proved to be remarkably accurate. Eratosthenes was in charge of a great scientific library at Alexandria, in Egypt, and from one of the books available to him he learned that at the time of the summer solstice, the "longest day" in northern latitudes, the Sun was vertically overhead at noon as seen from the town of Syene (the modern Assouan), some distance up the Nile. At Alexandria, however, the Sun was at this moment 7 degrees away from the overhead point, as is shown in Fig 2.2. A full circle contains 360 degrees, and 7 is about 1/50 of 360, so that if the Earth is spherical its circumference must be 50 times the distance from Alexandria to Syene. Eratosthenes may have arrived at the final figure of 24,850 miles, which is only fifty miles too small.*

If the Greeks had taken one step more, and placed the Sun in the centre of the planetary system, the progress of astronomy would have been rapid. Some of the philosophers tried to do so; but unfortunately Aristotle held the Earth to be the centre of the universe, and Aristotle's authority was so great that few people dared to question it. Moreover, the decentralization of the Earth would have meant a change in the laws of "physics", since Aristotle's idea of "things seeking their natural place" would have been much disturbed.

Most of our knowledge of Greek astronomy is due to Claudius Ptolemaeus (Ptolemy), who around AD 150 wrote a great book known generally by its Arab title of the *Almagest*. In it, he sums up the ideas of the great philosophers who had lived before him; and the theory that the Earth lies at rest in the centre of the universe is therefore called the "Ptolemaic", though as a matter of fact Ptolemy himself was not directly responsible for it.

On the Ptolemaic theory, all the celestial bodies move round the Earth. Closest to us is the Moon; then come Mercury, Venus, the Sun, Mars, Jupiter, Saturn and finally the stars. Ptolemy maintained that since the circle is the "perfect" form, and nothing short of perfection can be allowed in the heavens all these bodies must move in circular paths. Unfortunately, the planets have their own ways of behaving.

---

*There is some doubt whether Eratosthenes' estimate was accurate to within a few tens of miles, but at least his results were not wildly in error.

Ptolemy was an excellent mathematician, and he knew quite well that the planetary motions cannot be explained on the hypothesis of uniform circular motion round a central Earth. He therefore worked out a complex system according to which each planet moved in a small circle or "epicycle", the centre of which itself moved round the Earth in a perfect circle. As more and more irregularities came to light, more and more epicycles had to be introduced, until the whole system became hopelessly artificial and cumbersome.

Hipparchus, who had lived some two centuries before Ptolemy, had drawn up a detailed and accurate star catalogue. The original had been lost, but fortunately Ptolemy reproduced it in his *Almagest*, so that most of the work has come down to us. Hipparchus was also the inventor of an entirely new branch of mathematics, known to us as trigonometry.

When the power of Greece crumbled away, astronomical progress came to an abrupt halt. The great library at Alexandria was looted and burned in A.D. 640, by order of the Arab caliph Omar, though in fact most of the books may have been scattered earlier; in any case, the loss of the Library books was irreparable, and scholars have never ceased to regret it. For several centuries very little was done. When interest in the skies did return, it came – ironically enough – by way of astrology.

Even today, there are still some people who do not know the difference between astrology and astronomy. Actually, the two are utterly different. Astronomy is an exact science; astrology is a relic of the past, and there is no scientific basis for it, though in some countries (notably India) it still has a considerable following.

The best way to define astrology is to say that it is the superstition of the stars. Each celestial body is supposed to have a definite influence upon the character and destiny of each human being, and by casting a horoscope, which is basically a chart of the positions of the planets at the time of the subject's birth, an astrologer claims to be able to foretell the destiny of the person for whom the horoscope is cast. There may have been some excuse for this sort of thing in the Dark Ages, but there is none today. The best that can be said of astrology is that it is fairly harmless so long as it is confined to circus tents and the less serious columns of the Sunday newspapers.

However, mediaeval astrology did at least lead to a revival of true astronomy. The Arabs led the way, and presently interest spread to Europe. Star catalogues were improved, and the movements of the Moon and planets were re-examined. There were even observatories; very different from the domed buildings of today, but observatories nonetheless.

Astronomy was still crippled by the blind faith in Ptolemy's system. So long as men refused to believe that the Earth could be in motion, no real progress could be made. The situation was not improved by the attitude of the Church, which in those times was all-powerful. Any criticism of Aristotle was regarded as heresy. Since the usual fate of a heretic was to be burned at the stake, it was clearly unwise to be too candid.

The first serious signs of the approaching struggle came in 1546, with the publication of *De Revolutionibus Orbium Caelestium* (Concerning the Revolution of the Heavenly Bodies) by a Polish canon, Nicolas Copernicus. Copernicus was a clear thinker, as well as being a skilful mathematician, and at a fairly early stage in his career he saw so many weak links in the Ptolemaic system that he felt bound to abandon it. It seemed unreasonable to suppose that the stars could circle the Earth once a day. In his own words, "Why should we hesitate to grant the Earth a motion natural and corresponding to its spherical form? And why are we not willing to acknowledge that the *appearance* of a daily rotation belongs to the heavens, its *actuality* to the Earth? The relation is similar to that of which Virgil's Aeneas said, 'We sail out of the harbour, and the countries and cities recede.'"

Copernicus' next step was even bolder. He saw that the movements of the Sun, Moon and planets could not be explained by the old system even when all Ptolemy's circles and epicycles had been allowed for, and so he rejected the whole theory. He placed the Sun in the centre of the system, and reduced the status of the Earth to that of a perfectly ordinary planet.

Copernicus was wise enough to be cautious. He knew that he was certain to be accused of heresy, and though his book was probably complete by 1530 he refused to publish it until the year of his death. As he has foreseen, the Church was openly hostile. Bitter arguments raged throughout the next half-century, and one philosopher, Giordano Bruno, was burned in Rome because he insisted that Copernicus had been right.*

Tycho Brahe, born in Denmark only a few months after Copernicus died, was utterly unlike the gentle, learned Polish mathematician. Tycho was a firm believer in astrology, and an equally firm disbeliever in the Copernican system, so that it is ironical to realize that his own work did much to prove the truth of the new ideas. He built an observatory on the island of Hven, in the Baltic, and between 1576 and 1596 he made thousands of very accurate observations of the positions of the stars and planets, finally producing a catalogue that was far better than Ptolemy's. Of course, he had no telescopes; but his measuring instruments were the best of their time, and Tycho himself was a magnificent observer.

The story of his life would need a complete book to itself. Tycho is, indeed, one of the most fascinating characters in the history of astronomy. He was proud, imperious and grasping, with a wonderful sense of his own importance; he was also landlord of Hven, and the islanders had little cause to love him. His observatory was even equipped with a prison, while his retinue is said to have included a pet dwarf. Yet despite all his shortcomings, he must rank with the intellectual giants of his age. Nowadays, nothing remains of his great Uraniborg observatory.

When Tycho died, in 1601, he left his observations to his assistant, a young German mathematician named Johann Kepler. After years of careful study, Kepler saw that the movements of the planets could be explained neither by circular motion round the Earth, nor by circular motion round the Sun, so that there was something wrong with Copernicus' system as well as with that of Ptolemy. Finally, he found the answer. The planets do indeed revolve round the Sun, but not in perfect circles. Their paths, or "orbits", are elliptical.

One way to draw an ellipse is shown in Fig 2.3. Fix two pins in a board, and join them with a thread, leaving a certain amount of slack. Now loop a pencil to the thread, and draw it round the pins, keeping the thread tight. The result will be an

---

*This was not Bruno's only crime in the eye's of the Church, but it was certainly a serious one.

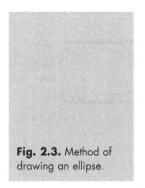

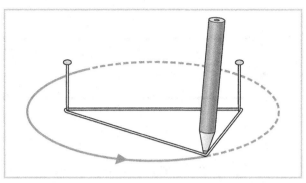

Fig. 2.3. Method of drawing an ellipse.

ellipse,* and the distance between the two pins or "foci" will be a measure of the eccentricity of the ellipse. If the foci are close together, the eccentricity will be small, and the ellipse very little different from a circle, if the foci are widely separated, the ellipse will be long and narrow.

The five planets known in Kepler's day proved to have paths which were almost circular, *but not quite.* The slight departure from perfect circularity made all the difference, and Tycho's observations fell beautifully into place, like the last pieces of a jig-saw puzzle. The age-old problem had been solved, though the Church authorities continued to oppose the truth for some time longer. Kepler's three Laws of Planetary Motion, the last of which was published in 1618, paved the way for the later work of Sir Isaac Newton.

Kepler's work was not the only important development to enrich the early part of the seventeenth century. In 1608 a spectacle-maker of Middelburg in Holland, Hans Lippershey, found that by arranging two lenses in a particular way he could obtain magnified views of distant objects. Spectacles had been in use for some time – according to some authorities, they were invented by Roger Bacon – but nobody had hit upon the principle of the telescope until Lippershey did so, more or less by accident.

A refracting telescope consists basically of two lenses. One, the larger, is the object-glass; its function is to collect the rays of light coming from a distant object, and bunch them together to form an image at the focus (Fig 2.4.). The image is then magnified by a smaller lens known as an eyepiece. This is more or less the principle used in the naval and hand telescopes of today, as well as in ordinary binoculars.

The news of the discovery spread across Europe, and came to the ears of Galileo Galilei, Professor of Mathematics at the University of Padua. Galileo, was quick to see that the telescope could be put to astronomical use, and "sparing neither trouble nor expense", as he himself wrote, he built an instrument of his own. It was a tiny thing, pitifully feeble compared with a modern pocket telescope, but it helped towards a complete revolution in scientific thought.

Galileo's first telescopic views of the heavens were obtained towards the end of 1609. At once, the universe began to unfold before his eyes. The Moon was covered

*The method is excellent in theory. In practice, what usually happens is that the pins fall down or the thread breaks. One day, I hope to carry out the whole manoeuvre successfully.

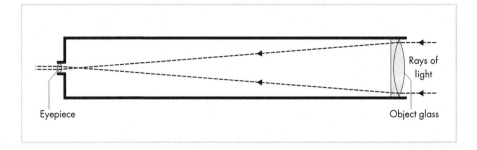

Fig. 2.4. Principle of the refractor.

with dark plains, lofty mountains and giant craters; Venus, the Evening Star of the ancients, presented lunar-type phases, to that it was sometimes crescent, sometimes half and sometimes nearly full; Jupiter was attended by four moons of its own, while the Milky Way proved to be made up of innumerable faint stars.

Galileo had always believed in the new system of the universe, and his telescope work made him even more certain. Inevitably he found himself in trouble with the Church. It was hard for religious leaders to realize that the Earth is not the most important body in the universe, and Galileo seemed to them to be a dangerous heretic. He was arrested and imprisoned, after which he was brought to trial and forced to "curse and abjure and detest" the false theory that the Earth moves round the Sun.

Few people were deceived, and before the end of the century the Ptolemaic theory had been abandoned for ever. The publication of Isaac Newton's *Principia*, in 1687, led to real understanding of the way in which the planets move.

Most people have heard the story of Newton and the apple. It is interesting because unlike most stories of similar type, such as Canute and the waves, it is probably true. Apparently Newton was sitting in his garden one day when he saw an apple fall from its branch to the ground, and upon reflection he realized that the force pulling on the apple was the same force as that which keeps the Moon in its path round the Earth. From this he was led on to the idea of "gravitation", upon which the whole of later research has been based. It is fair to say that Kepler found out "how" the planets move; Newton discovered "why" they do so.

Newton also constructed an entirely new type of telescope. As has been shown, Galileo's instrument was a refractor, and used an object-glass to collect its light. Newton came to the conclusion that refractors would never be really satisfactory, and looked for some way out of the difficulty. Finally he decided to do away with object-glasses altogether, and to collect the light by means of a specially-shaped mirror.

When Newton rejected the refractor as unsatisfactory, he was making one of his rare mistakes. However the Newtonian "reflector" soon became popular, and has remained so. Mirrors are easier to build than lenses, and even today all the world's largest instruments are of the reflecting type.

Astronomy was growing up. So long as observations had to be made with the naked eye alone, little could be learned about the nature of the planets and stars; their movements could be studied, but that was all. As soon as telescopes became available, true observatories made their appearance. Copenhagen and Leyden took the lead; the Paris Observatory was completed in 1671, and Greenwich in 1675.

Greenwich was founded for a special reason. England has always been a seafaring nation, and before the development of reliable clocks the only way in which sailors could fix their position when far out in the ocean, out of sight of land, was to observe the position of the Moon among the stars. This involved the use of a good star catalogue, and the best one available, Tycho's, was still not accurate enough. Charles II therefore ordered that the star places must be "anew observed, examined and corrected for use of my seamen". A site was selected in the Royal Park at Greenwich, and Sir Christopher Wren, himself a former professor of astronomy, designed the first observatory building. The Rev John Flamsteed was appointed Astronomer Royal, and in due course the revised star catalogue was completed.

Telescopes continued to be improved. Some of the early instruments were curious indeed; one of them, used by the Dutch observer Christiaan Huygens, was over 200 feet long, so that the object-glass had to be fixed to a mast. But gradually the worst difficulties were overcome, and both refractors and reflectors gained in power and in convenience. Mathematical astronomy made equally rapid strides. The great obstacle had always been the Ptolemaic system, and once that had been swept away the path was clear. The distance between the Earth and the Sun was measured with fair accuracy, and in 1675 the Danish astronomer Ole Rømer even measured the speed of light, which proved to be 186,000 miles per second. Rømer did this, incidentally, by observing the movements of the four bright moons of Jupiter.

But though knowledge of the bodies of the Solar System had improved out of all recognition, little was known about the stars, which were still regarded as mere points of reference. The first serious attack on their problems was made by William Herschel, who is rightly termed the "father of stellar astronomy".

Herschel was born in Hanover in 1738, eleven years after the death of Newton. He came to England, and became organist at the Octagon Chapel in Bath; but his main interest was astronomy, and he built reflecting telescopes which were the best of their age. The largest of Herschel's telescopes, completed in 1789, had a mirror 48 inches in diameter and a focal length of 40 feet. The mirror still exists, and now hangs on the wall of Flamsteed House in Greenwich, though it has not been used since Herschel's time.

Herschel had his living to earn, and for some years he could not afford to spend all his time in studying astronomy. Then, in 1781, he made a discovery which altered his whole life. One night he was examining some faint stars in the constellation of the Twins, when he came across an object which was certainly not a star. At first he took it for a comet, but as soon as its path was worked out there could no longer be any doubt as to its nature. It was not a comet, but a planet – the world we now call Uranus.

The discovery was quite unexpected. There were five known planets, and these, together with the Sun and Moon, made a grand total of seven. Seven was the magical number of the ancients, and it had therefore been thought that the Solar System must be complete. Herschel became world-famous; he was appointed Court Astronomer to King George III, and henceforth he was able to give up his musical career altogether.

Herschel set himself a tremendous programme. He decided to explore the whole heavens, so that he could form some idea of the way in which stars were arranged. Until the end of his long life, in 1822, he worked patiently at his task, and his final conclusions have been proved to be reasonably accurate.

Naturally, Herschel made numerous discoveries during his sky-sweeps. Many apparently single stars proved to be double, and there were also clusters of stars, as well as faint luminous patches known as "nebulae", from the Latin word meaning "clouds". Herschel was a most painstaking observer. He catalogued all his discoveries, and when we examine his published papers we can only marvel at the amount of work he managed to do. Since he lived in England for most of his life, he was unable to examine the stars of the far south, which never rise in northern latitudes, and it was fitting that the completion of his sky-sweeps should be accomplished later by his son, Sir John Herschel, who travelled to the Cape of Good Hope specially for this purpose, and remained there for several years.

Another famous observer of this period was Johann Schröter, chief magistrate of the little German town of Lilienthal. Unlike Herschel, Schröter concentrated mainly upon the Moon and planets, and he is the real founder of "selenography", the physical study of the lunar surface. Unfortunately Schröter's observatory, together with all his unpublished work, was destroyed by the invading French armies in 1814, and Schröter himself died two years later.

In the early years of the nineteenth century a German optician, Fraunhofer, began to experiment with glass prisms. Newton had already found that ordinary "white" light is not white at all, but is a blend of all the colours of the rainbow. Fraunhofer realized that this discovery could be turned to good account, and his work led to the development of a new instrument, the astronomical spectroscope.

Just as a telescope collects light, so a spectroscope analyses it. By studying the "spectra" produced, it is possible to find out a great deal about the matter present in the material which is emitting the light. For instance, the spectrum of the Sun shows two dark lines which can be due only to the element sodium, so that we can prove that sodium exists in the Sun.

Today we can examine the spectra of stars and star-systems so far away that their light takes thousands of millions of years to reach us – and we find the same familiar elements.

In 1838, Friedrich Bessel, Director of the Observatory of Köningsberg, returned to the problem of the distances of the stars. By studying the apparent movements of 61 Cygni, a faint object in the constellation of the Swan, he was able to show that it lay at a distance of about 60 million million miles. About the same time a British astronomer, Henderson, measured the distance of the bright southern star Alpha Centauri, and arrived at the reasonably accurate value of twenty million million miles; the real value is about 24 million million miles, so that Henderson underestimated somewhat. Alpha Centauri is a triple star, and the faintest member of the trio remains the nearest known body outside our own Solar System.

Twenty four million million miles! Our brains are not built to understand such vast distances, and it is clear that the mile is too short to be a convenient unit of length. One might as well try to measure the distance between London and Melbourne in centimetres. Fortunately there is a much better unit available, based upon the speed of light.

Light is known to travel at 186,000 miles per second. A ray from the Sun takes $8\frac{1}{3}$ minutes to reach us, but in the case of Alpha Centauri the time of travel is $4\frac{1}{3}$ years; we see the star not as it is now, but as it was $4\frac{1}{3}$ years ago. Alpha Centauri is therefore said to be $4\frac{1}{3}$ light-years away, while the distance of 61 Cygni is nearly 11 light-years.

Bessel's success gives us an added idea of the real importance of the Solar System. Rather than quote strings of figures, it will be more graphic to imagine a scale model. If we begin with making the Sun a 2-foot globe, and putting it on Westminster Bridge, the Earth will become a pea at a distance of 215 feet; Uranus, the outermost of the planets known in Bessel's time, will be represented by a plum $\frac{4}{5}$ of a mile away from our 2-foot Sun. What of the nearest star? We shall not find it in London, or even in England; it will lie some 10,000 miles away, in the frozen wastes of Siberia. We have learned much since the days when the Earth was thought to be the hub of the universe.

A vital development in the early 19th century was the beginning of astronomical photography. The first "Daguerreotype" picture of the Sun was taken in 1845, followed in 1850 by a good photograph of the Moon. Progress was rapid, within fifty years magnificent photographs were being taken, not only at professional observatories but also by amateurs. Only since around 1960 has photography in its turn been superseded by electronic devices. Officially, sheer visual observation belongs to the past and it is seldom that a professional astronomer actually looks through a telescope. Times have indeed changed.

Herschel's 48-inch reflector was outmatched in 1845, when the third Earl of Rosse, at Birr Castle in Ireland, constructed a 72-inch. He built it himself – even grinding the mirror – and though it was awkward to use, and could examine only part of the sky, it was by far the most powerful telescope then in existence. Lord Rosse used it well. In particular, he paid close attention to the star-clusters and nebulae which had been pointed out by Herschel and catalogued by the French observer Charles Messier. Some of the nebulae proved to be made up of faint stars, but others could not be resolved. Many of the "starry nebulae" revealed a spiral structure, so that they looked like shining Catherine-wheels.

For some years the Rosse telescope was in a class of its own, but well before the end of the century the first large refractors were built, and were far more convenient and effective. In fact, the 72-inch fell into disuse after 1909, but it has now been restored, though its main interest is historical. Nothing like it has been built either before or since, but it was a great achievement, and will always be remembered for its discovery of the spirals.

Alone, the telescope could never decide upon the nature of the irresolvable nebulae; the spectroscope was able to do so. In 1864 Sir William Huggins examined a faint nebula in the Dragon, and found that it was made up not of stars, but of luminous gas.

It is now known that the nebular objects are of three types, Inside our own star-system, known commonly as the Milky Way but more properly as the Galaxy, we find the normal star-clusters and the gaseous nebulae, most of them hundreds or thousands of light-years from us. Beyond the Galaxy there is a vast gulf, and then we come to the separate external systems, lying at immense distances. The most famous of them is the Great Spiral in Andromeda (Fig. 2.5), which can be seen with the naked eye as a faint misty patch, and which proves to be a galaxy in its own right, even larger than our own. Herschel had suspected something of the sort, and the work of Rosse and Huggins supported his view, though the question was not finally settled until 1923.

The latter part of the nineteenth century was the age of the great refractors. The largest, set up in 1897 at the Yerkes Observatory at Williams Bay in Wisconsin, has an object-glass 40 inches in diameter. It is unlikely to be surpassed, at least so far as Earth-based telescopes are concerned, because a lens has to be supported around its edge, and if it becomes too heavy it tends to distort, making it useless. Other big refractors still in use are those of the Lick Observatory in California (36 inches), Meudon in France (33) and the Lowell telescope at Flagstaff in Arizona (24). Reflectors then took over, largely due to the energy of one American astronomer, George Ellery Hale, who not only planned large reflectors but had the happy knack of persuading millionaires to finance them. He was responsible for the 60-inch reflector at Mount Wilson, California, and then in 1917, the 100-inch Hooker reflector, also at Mount Wilson.

Fig. 2.5. The Andromeda Galaxy – like our Galaxy, a spiral system. (Photographed by Bill Patterson with his Takahasi FSQ106 telescope and SBIG ST10XME CCD camera. See www.laastro.com.)

For decades the 100-inch was supreme; only it was powerful enough to enable Edwin Hubble to prove that the spirals seen from Birr were independent galaxies, millions of light-years away, rather than minor features of our own Milky Way. It was finally surpassed in 1947 by the 200-inch reflector at Paloma, in California. Others have followed, some with huge mirrors made up of many smaller components fixed together to form the correct optical curve ("segmented mirrors"), while in many cases there are large telescopes working together. Such are the Keck twins, on Hawaii, each of which has a 387-inch (10-metre) mirror. At the moment pride of place goes to the Very Large Telescope in Northern Chile. There are four 8-metre mirrors which can be used either individually or collectively. In theory, the VLT could detect the headlights of a car, separately, over a range of more than 20,000 miles.

Originally, most major observatories were in the northern hemisphere, but it so happens that some of the most important stellar objects lie in the far south of the sky, and recent emphasis has been on places such as Chile, Australia and South Africa. It is also important to consider conditions of "seeing"; the less atmosphere you have above, the better the conditions are likely to be. This is why major telescopes have been located at, for instance, Mauna Kea in Hawaii, at an altitude of 14,000 feet, and at the lofty Atacama Desert of Chile. Britain's largest telescope, the 165-inch William Herschel reflector, has been set up atop an extinct volcano, the Roque de los Muchachos, in Las Palma, one of the Canary Islands.

Obviously there are tremendous advantages in going into space, above the atmosphere, and in 1990 the Hubble Space Telescope was launched into orbit. Moving round the Earth at an altitude of over 370 miles it can in some ways outmatch any telescope on the ground, even though it has a mirror "only" 94 inches in diameter.

There are many optical systems in use; for example a Schmidt telescope can photograph a comparatively wide area of the sky with a single exposure. However, it is fair to say that electronic devices are now taking over from photography, and many modern amateurs are extremely well equipped. With a CCD or Charge-Coupled Device, a 12-inch telescope can match the performance of, say, the Mount Wilson 60-inch used with conventional film. Nowadays it is seldom that a professional astronomer actually looks through a telescope; instead he sits in a warm, comfortable room studying the results as they come through on a television screen.

Radio astronomy began in the early 1930s, when an engineer named Karl Janksy, working for the Bell Telephone Company, was investigating problems of "static" and found that he was picking up radio waves from the sky. This was the beginning of radio astronomy, which has now come so very much to the fore.

Radio telescopes are not in the least like optical telescopes, and they do not produce visible pictures of the objects under study; one cannot look through them, as some earnest inquirers fondly believe! They are designed to collect the long-wavelength radiations coming from space, and they are of many different designs. The most famous radio telescope is probably the 250-foot steerable "dish" at Jodrell Bank, in England (Fig. 2.6), but each design is tailored to suit its own special needs. I am not a radio astronomer, but electronically-minded amateurs will certainly find plenty of scope. Grote Reber, who built a "dish" before the war and was probably the first true radio astronomer, was an amateur.

Fig. 2.6. Jodrell Bank Radio Telescope, Cheshire. (Aerial photograph by Jonathan C.K. Webb.)

The opening of the Space Age caused a complete change in outlook. On 4 October 1957 the Russians launched the first space satellite, Sputnik; since then men have been to the Moon, unmanned probes have explored all the planets except Pluto, and instruments have been landed not only on Mars and Venus but also upon the tiny asteroid Eros. Yet all this does not mean that the day of the amateur astronomer is over. His field of research is much more restricted than it used to be, but his role remains as important as ever.

I am writing these words in early 2005. Before they appear in print, much may have happened, but the basic problems will remain unaltered. And as one puzzle is solved, a host of others arises to take its place. This has been the case since ancient times; it is still the case today.

Chapter 3

Telescopes and Observatories

It is often thought anyone aspiring to take up astronomy as a hobby should rush straight out and buy a telescope. In fact, nothing could be further from the truth. There is a great deal within the range of the naked-eye observer, and binoculars are invaluable. They will not show spiral galaxies, the rings of Saturn, or the ice caps of Mars, but they will give endless enjoyment, and I have often said that I would prefer a good pair of binoculars to a very small telescope.

However, sooner or later the newcomer will feel the need for a telescope, and obviously this means spending a fair sum of money. It's also true that everything depends upon the main interests of the observer. But before going any further, let me say something about the various types of telescopes, starting with the refractor (a pair of binoculars is made up of two small refractors working together). The basic principle is straightforward. The rays of light coming from the object under observation are collected by a lens or object-glass, which bunches the rays together and brings them to focus. The image produced is then enlarged by another lens, known as the eyepiece. All the actual magnification is done by the eyepiece, and various eyepieces can be fitted to the same telescope.

This seems simple enough, but there are complications. For instance, the eyepiece is generally not a single lens, but a group of lenses held in a casing. The final view will be upside-down, unless deliberately corrected, but this does not matter in the least; in most astronomical photographs and drawings the south is at the top of the picture, with west to the left.

Even the object-glass is not a single lens, and the reason for this is rather interesting. As Newton discovered, what we call "white" light is made up of all the colours of the rainbow, from red to violet. Light may be considered as a wave motion, and the distance from one crest to the next is called the wavelength (Fig. 3.1). Red light has a longer wavelength than blue or violet, and the result is that the object-glass does not bend it so much. The difference in the amount of bending or "refraction" means that the red rays are brought to focus at a greater distance from the object-glass (Fig. 3.2). This causes trouble, and the image of a bright object will appear to be surrounded by false colour.

Newton failed to find the remedy, and it was for this reason that he abandoned refractors altogether. Actually, there is at least a partial answer. Modern object-glasses are made up of several lenses, composed of different kinds of glass whose

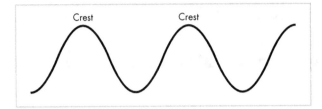

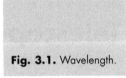

Fig. 3.1. Wavelength.

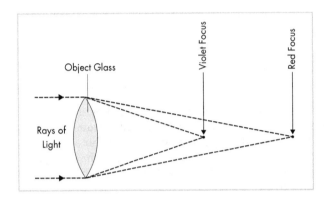

Fig. 3.2. Unequal refraction. The difference between the refraction of red and violet light has been very much exaggerated, for the sake of clarity.

chromatic properties tend to lessen the trouble. The effect can never be eliminated, but it can be very much reduced.

A refractor is classified by the diameter of its object-glass. A "3-inch" has an object-glass 3 inches across, and so on. The average amateur-owned refractor will have an object-glass between 3 and 12 inches across; anything larger than that is decidedly unusual.

The distance between a lens and its focal point is known as its "focal length", and this length divided by the diameter of the object-glass gives the "focal ratio" (usually abbreviated to "f/ratio"). For instance, I have a 3-inch refractor with a focal length of 36 inches. The f/ratio is therefore 36 ÷ 3, or 12. The eyepiece combination has its own focal length, and the magnification obtained depends on the ratio of the focal length of the eyepiece to that of the object-glass. In the case of my own $\frac{f}{12}$ refractor, an eyepiece of focal length $\frac{1}{2}$ inch will give a magnification of 36 ÷ $\frac{1}{2}$, or 72 diameters – usually written, for short, as "×72". With an object-glass of focal length 48 inches, the same eyepiece would give a power of 48 ÷ $\frac{1}{2}$, or ×96.

It might therefore be thought that the way to get the best out of an eyepiece would be to use it with an object-glass of long focal length. Unfortunately this introduces other troubles, and the only solution is to strike a happy mean.

Naturally, a large object-glass will collect more light than a smaller one. Suppose that I use a very short-focus eyepiece, say $\frac{1}{20}$ inch, upon my 3-inch refractor? The magnification will be 36 ÷ $\frac{1}{20}$, or 720. Yet the image will be so faint that nothing will be made out. The small object-glass, only 3 inches across, is quite unable to collect enough light to satisfy so powerful an eyepiece. If I want to use a magnification of 720, I must buy a larger telescope.

Lens-making is a tricky business, and not many amateurs will feel inclined to tackle it (certainly I would not). If you want a refractor, you will have to buy it

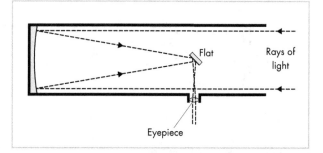

Fig. 3.3. Principle of the Newtonian reflector. For the sake of clarity, the curve of the main mirror has been much exaggerated.

Flat

Rays of light

Eyepiece

ready-made. So let us next consider the reflector, and here we have a number of different optical systems.

Newton's arrangement is shown in Fig. 3.3. Here the light from the distant object passes straight down an open tube until it strikes a mirror at the bottom. This mirror is shaped so as to reflect the rays back up the tube, directing them on to a smaller mirror called a flat. The flat is placed at an angle, and sends the rays to the side of the tube, where they are brought to focus and are magnified by an eyepiece in the ordinary way. With a Newtonian reflector, therefore, the observer looks into the side of the tube instead of up it. Of course, the flat prevents some of the light-rays from reaching the main mirror at all, but the loss is not serious, and in any case there is no way of avoiding it.

There is one great advantage in getting rid of the object-glass. A mirror reflects all colours equally, and so the troublesome colour fringes do not appear. For this reason, colour estimates with a reflector are a good deal more reliable than those made with the help of a refractor.

A reflector is classified according to the diameter of its main mirror. However, we must be careful when comparing mirrors with lenses; inch for inch, the lens will give a better result. A 6-inch refractor is appreciably more effective than a 6-inch reflector, so that it can be used with an eyepiece of higher magnification.

Generally speaking, small and moderate reflectors have focal ratios of from f/7 to f/9. There are good reasons for this, but to enter into a full discussion would be beyond our present scope. Nor need we do more than mention the other types of reflecting telescopes; the Gregorian and the Cassegrain, in which the light is reflected back through a hole in the main mirror (Fig. 3.4), and the Herschelian, in which the main mirror is tilted, ao as to dispense with the flat altogether (Fig. 3.5). The Herschelian may sound attractive, because there is no flat to block out part of the main mirror, but there are horrible disadvantages, which need not concern us here, and few Herschelians are in use today. I have yet to look through one. On the other hand, the Cassegrain is becoming more and more popular, and has the advantage of being pleasingly compact.

The performance of a reflector depends entirely upon the quality of its mirror, which have to be amazingly accurate in form. Glass mirrors are coated with a thin layer of silver, aluminium or occasionally rhodium and have to be periodically re-polished, which is not so easy as might be thought.

"Catadioptric telescopes, involving both lenses and mirrors, are now very much in vogue – mainly Meades and Celestrons. (They are so named because catoptric

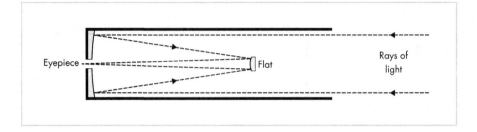

Fig. 3.4. Principle of the Cassegrain reflector. The "flat" is convex and is just in front of the point of focus of the main mirror. In the Gregorian reflector, the "flat" is concave and is placed just beyond the point of focus of the main mirror. The Gregorian gives an erect image.

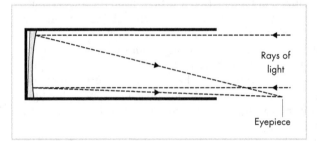

Fig. 3.5. Principle of the Herschelian reflector. The form of reflector is now virtually obsolete.

systems use only mirrors, and dioptric systems use only two lenses; the Meades and Celestrons are made by the two leading commercial firms.) Fig. 3.6 shows the principle of the Schmidt-Cassegrain. Telescopes of this type are efficient and relatively portable; neither do they need much maintenance. The main mirror is spherical, and there is a special correcting plate of complex shape. They used to be prohibitively expensive, but prices today are much more reasonable.

There was a time when amateur telescope-making was very popular, and anyone with reasonable skill could make an adequate reflector, including the main mirror. I have myself made a few mirrors, and I am probably one of the clumsiest people in the United Kingdom. So it may be as well to give a few brief comments, just in case anyone reading this book decides to make the attempt.

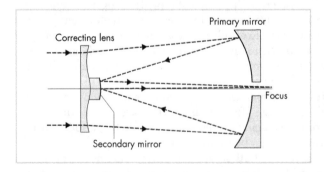

Fig. 3.6. Principle of a Schmidt-Cassegrain catadioptric telescope.

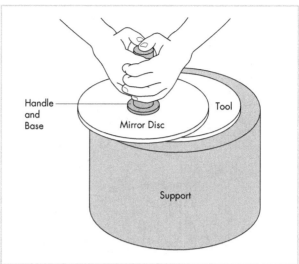

Fig. 3.7. Grinding a mirror.

The principle of grinding a mirror into the correct optical curve is to use two disks of glass, at least an inch thick, one which will turn into the final mirror, while the other is merely a "tool". The tool is fastened to a bench, and the mirror placed on top of it, with water and carborundum powder between the two. The mirror is then slid to and fro, while the operator rotates it and also walks round the bench. Clearly, the tool will be worn away round the edge and will thus become convex, while the mirror will be worn away in the middle and will thus became concave (Fig. 3.7).

This process is easy, and needs merely a good deal of patience until the curve is more or less correct. The mirror has then to be polished and figured, and a moment's carelessness will ruin hours of work. Numerous tests have to be made, and the real difficulty lies in the "figuring", which means producing, the correct curve. But it can be done, and making a 6- or 8-inch mirror is within the capabilities of most people. I know of a fifteen-year-old enthusiast who has made himself a really good 6-inch reflector, and has also built the stand. Remember, though, that nobody should set out to grind a mirror without being prepared for a series of setbacks. Difficulties and problems arise at every turn, and there will be moments when the luckless operator feels inclined to hurl his mirror on to the ground and stamp on it. Patience is absolutely necessary – as is the case with almost everything in life.

The construction of a mount is purely a mechanical task. One form, the altazimuth, is shown in Fig. 3.8. The instrument – in this case a 6-inch reflector – is resting in a cradle (A), and is kept in position solely by its own weight. The cradle can be rotated (B), and the telescope can be swung up or down by sliding the rod (C). The top of the rod is fitted with a worm (D), so that by moving the wheel the telescope can be moved very slightly up or down, while the handle (E), attached to a special form of joint, gives a similar slight rotation of the whole telescope. D and E are know as "slow motions". They are not essential, but they are certainly helpful.

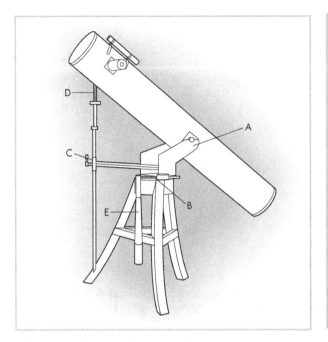

Fig. 3.8. Altazimuth mount, for a small reflector.

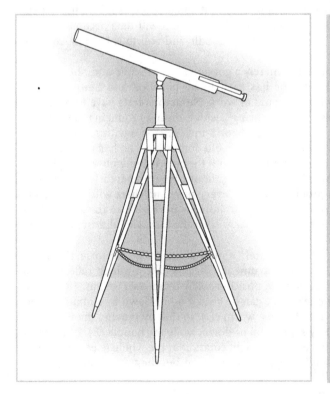

Fig. 3.9. Simple tripod mount for a small refractor.

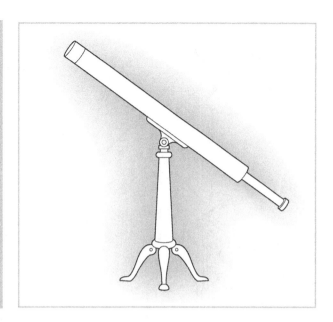

Fig. 3.10. Pillar and claw mount for a small refractor. I have nicknamed it the "Blancmange" mount, for reasons which should be obvious to anyone who has used it.

Fig. 3.9 shows a much simpler mount, this time for a 3-inch refractor. It is simply a tripod, so that the telescope can be moved in any direction; slow motions are not always fitted, but certainly make for easier observing.

The next drawing, Fig. 3.10, is included as an Awful Warning. It is that appalling contrivance known as the Pillar and Claw Stand, beloved of dealers and despised by serious amateurs. It looks nice, and it is cheap, but it is about as steady as a blancmange. The slightest puff of wind will cause the whole telescope to quiver, and the object under observation will appear to dance about like dice in a shaker. Anyone who buys a small refractor may well find that it is mounted upon a pillar and claw. If any real work is to be done, the only solution is to buy a rigid tripod and consign the original stand to the dustbin.

The Dobsonian mounting has become popular, and is very straightforward. The telescopes mounted in what is to all intents and purposes a box, so that it can swing up and down, the box can be rotated on a base. The Dobsonian mounting can carry large telescopes, and is of course portable. The main disadvantages is that movements are crude, and high magnifications are not easy to use.

Lastly, we come to the Equatorial Stand (Fig. 3.11), which is far better than any of those previously described. For a telescope of any size, an equatorial mounting is highly desirable, because the Earth is in rotation.

The spinning of the Earth from west to east means that all the celestial bodies appear to move from east to west. This movement is slow, judged by everyday standards, but when we use a telescope to magnify the size of an object in the sky we also magnify the apparent motion. If the telescope remains stationary, a star or planet will seem to shift steadily across the field until it disappears from view. The telescope has then to be moved until the object is found again. Moreover, there are

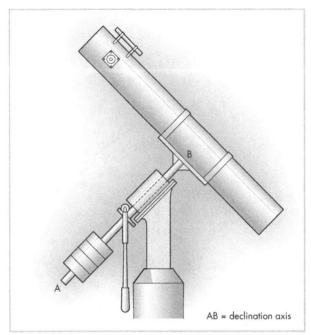

AB = declination axis

Fig. 3.11. Equatorial mount.

two motions to be made: up or down ("declination"), and east to west ("right ascension"). Slow motions of the type shown in Fig. 3.8 provide one answer, and are helpful, but it is irritating to have to fiddle continuously with both wheel and handle. To work in comfort under such conditions, one would need four or five hands.

In the equatorial stand, the "polar axis" is pointed towards the celestial pole, so that only the east-to-west pushing is necessary – the telescope will take care of the up-or-down motion of its own accord. If possible, a driving motor should be attached and regulated so that the telescope moves slowly round at a speed which compensates for the apparent shift of the celestial bodies across the sky.

Conventionally, the mirrors for a refractor are set in a tube, but can equally well be held in a box-like structure; after all, the only essential is to keep all the optical components in the correct position. The "tube" may even be a skeleton, provided that it is really firm.

One addition is simplicity itself. A small sighting telescope or "finder" can be fitted, and will be found most useful (Fig. 3.13). Even a toy telescope will do, and can be attached by Meccano. The advantage of a finder is that it has a large field of view, and will save much time when a faint object is being searched for. The object is simply brought to the centre of the finder field; if the adjustments are correct, the object will then be visible in the field of the main telescope.

Come now to the choice of a telescope – and it is fair to say that today the average amateur will prefer to buy a telescope rather than go to the trouble of making one for himself. Great care is needed, notably when buying a second-hand telescope, particularly a reflector.

Fig. 3.12. A fork mounting.

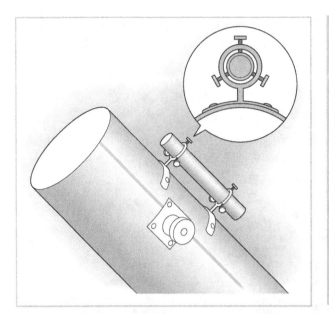

Fig. 3.13. Fitting a finder to a reflector.

Fig. 3.14. An example of a finder mounted on a telescope. (Photograph courtesy of Gordon Rogers.)

It may look perfectly sound and well mounted, but if the mirror is poor the performance also will be poor, and a bad mirror does not always betray itself at first glance. If you have no expert knowledge, my advice is to consult a member of your local Astronomical Society, who will certainly come to the rescue. Advertisements can be misleading. The function of the main mirror (or lens, with a refractor) is to collect light; the actual magnification is done by the eyepiece. If you see an advertisement for a telescope "magnifying × number of times", avoid it all costs. It is aperture which counts. The maximum usable magnification for a telescope is of the order of × 50 per inch of aperture; thus a 3-inch telescope will gear up to ×150, while a 12-inch will go up to × 600 under really good conditions. Not long ago I saw an advertisement for a telescope capable of giving a magnification of × 5000. It proved to be a 3-inch refractor, and could not possibly staff more than × 150 (Fig. 3.14).

Reflector, refractor or catadioptric? Refractors are easier to handle, and are not temperamental. I have always said that the minimum aperture for a useful refractor is 3 inches; I am bound to admit that over the past year or two some reasonable 2½-inch refractors have come on the market, and cost less than £100, but I would still prefer a 3-inch or larger, and the cost does not seem too great when compared with, for instance, a couple of train tickets between London and Manchester. A refractor of aperture larger than 4 inches will be only semi-portable. Incidentally, make sure that you have a dew-cap, which is simply a short tube which fits over the object-glass end of the refractor in order to prevent dust, dirt and dew from settling on the lens. It can be made from a cocoa-tin lined with blotting paper, or something of a kind, and a cap should always be kept over the object-glass when the telescope is not actually in use (Fig. 3.15).

If the object-glass needs cleaning, it should be brushed very gently with a camel's-hair brush and then wiped even more gently with a piece of very fine, clean silk or wash-leather. To take the various components of an object-glass apart is most unwise unless the owner has a really good idea of what he is about. All things considered, a small refractor should need little or no attention for years on end, provided that it is not roughly handled. When some major adjustment does become necessary, it will be worth while to take the whole instrument to an expert. It is better to spend a little money on maintenance than a great deal of money on buying a new telescope.

Reflectors need more attention. The main mirror and the flat need periodical re-silvering, and although this can be done at home it does need a good deal of care.

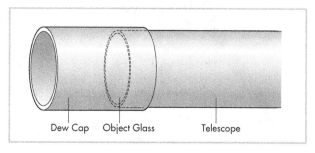

Fig. 3.15. Dew-cap for a refractor.

Dew Cap   Object Glass   Telescope

It is probably better to have the mirror aluminized, which will give a much longer period before anything further need be done; rhodium coating can also be used. Both mirror and flat should be kept covered with a protecting cap except when actually in use, and yet another word of warning may be timely here. Before using the telescope, uncap the flat before you expose the main mirror. I know of one luckless observer who uncovered the main mirror first – and then dropped the flat cover on to it. He spent the next few months grinding himself a new mirror.

Eyepieces are vitally important, since using a good telescope with a bad eyepiece is like using a good record-player with a bad needle. Theoretically (though not always in practice) eyepieces are made to a standard thread, so that any eyepiece should fit any telescope; but the magnification obtained depends upon the focal length of the mirror or object-glass, so that an eyepiece which yields × 50 on a 3-inch refractor will not yield × 50 on a 6-inch. Moreover, eyepieces are of various types, adapted for different types of telescopes.

It is advisable to have at least three eyepieces. One should give low magnification, for star-sweeping and general views; the second, moderate magnification for more detailed views of planets and some stellar objects; the third, high magnification for use on really good nights.

Fig. 3.16. A Celestron telescope.

Never try to use too high a power. If the image becomes even slightly blurred, change at once to a lower magnification. It may be impressive to say that an observation was made "× 400" or "× 500", but it will often be found that a smaller, sharper picture will yield far more detail.

All in all, there is little doubt that a catadioptric telescope – a Meade or a Celestron – is the ideal instrument for the modern amateur. It may well be computerized, so that it will locate selected targets automatically, it will be portable if need be, it will be versatile, and it will need little maintenance provided that it is properly looked after (Fig. 3.16). The 6-inch Meade shown in the photograph cost £499 when I bought it in 1999. No doubt prices will go up, but always remember that the cost is non-recurring. The telescope will last you a lifetime.

What, then, about observing sites, particularly at the present time, when light pollution has become such a menace?

A favourite mistake is to poke a telescope through the bedroom window in the expectation of seeing fine detail on the Moon or a planet. Actually, good results can seldom or never be obtained in such a way. The temperature of the room is almost certain to be higher than that outside, and the resultant air-currents will destroy the sharpness of the image. Moreover, there are other hazards. A friend of mine once dropped a precious eyepiece fifteen feet on to a gravel path, with predictable results. The really dedicated amateur may well want an observatory, and if a suit-

Fig. 3.17. The runoff shed for my $12\frac{1}{2}$-inch reflector. (Photo taken by Alan Schultz in May 2004).

able site is available this is an excellent solution. (Note that an observatory standing on a concrete base, and not fastened down, is technically a portable structure, and is not subject to planning permission. It is useful to bear this in mind if you have trouble with an awkward local council.)

A run-off observatory is easy to build; it runs on rails, so that when the telescope is to be used the shed can be rolled back out of the way. It is best to have the shed in two halves, rolling back in opposite directions; a single section involves a door, which tends to flap and may act as a powerful sail when the observer is trying to replace it under windy conditions. The main disadvantage of a run-off shed is that it gives no protection against either wind or stray light.

A run-off roof arrangement is another useful pattern, though better suited to a refractor than a reflector. Of course, the classic observatory is domed, and this is ideal, because it protects the telescope completely and also shields the observer. Be careful about the choice of site – and it always happens that your neighbour's high tree lies in the most inconvenient position possible. If you want to build an observatory, the books listed in the Appendix will show you how to do it. I wish you all success (Figs 3.17 and 3.18).

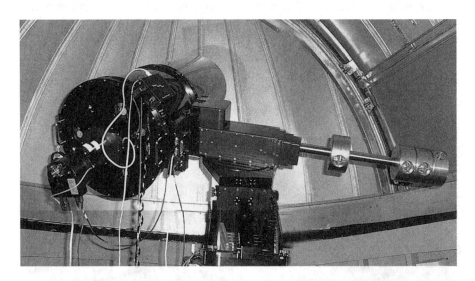

Fig. 3.18. Gordon Rogers' 16-inch telescope at Long Crendon Observatory, Oxfordshire.

Chapter 4

The Solar System

Some people refuse to take an interest in astronomy simply because they are frightened of it. They cannot appreciate distances of millions of miles; they cannot believe that each star is a sun, and their minds remain firmly anchored to the Earth.

The point of view is commoner than might be imagined, and part of the difficulty originates from the vast scale of the universe. Nobody can really picture "a million miles", and the tremendous heat of the Sun's interior is equally beyond the human brain. The best way to give some account of scale is to visualize a model, which will at least put our ideas in some sort of order.

The Solar System in which we live is made up of one star (the Sun), nine major planets, and numerous bodies of lesser importance, such as the moons or "satellites", the minor planets, the meteors and the comets. Returning to the model discussed on page 30, we imagine that the Sun has become a globe only 2 feet in diameter, so that we can put in the rest of the planets on the correct scale. Mercury will become a grain of mustard seed 83 feet from a 2-foot Sun; Venus, a pea at 156 feet; the Earth; another pea at 215 feet; Mars, a pin's head at 328 feet; Jupiter, an orange at $\frac{1}{5}$ of a mile; Saturn, a tangerine at $\frac{2}{5}$ of a mile; Uranus, a plum at $\frac{4}{5}$ of a mile; Neptune, another plum at $1\frac{1}{4}$ miles; and Pluto, another seed grain with a maximum distance of 2 miles. The nearest of the ordinary stars will then lie 10,000 miles off, which gives us a good idea of how isolated the Solar System really is.

There is a great deal of difference between a 2-foot globe and an orange, and so even Jupiter, largest and most massive of the nine planets, is by far inferior to the Sun. The Sun is in fact the absolute ruler of our system; it controls the movements of the planets, and the planets depend entirely upon solar heat and warmth. No planet has any light of its own. Even Venus, the glorious "evening star" which can shine down like a small lamp and can even cast a shadow at times, is in itself a non-luminous body.

One thing is evident from our scale model: the planets can be divided into two well-marked groups. The inner group is made up of four small and comparatively close-in worlds, Mercury, Venus, the Earth and Mars. Then comes a wide gap, followed by the four giants, with that curious little world Pluto on the very fringe of the Sun's kingdom. Actually, the gulf between Mars and Jupiter is not empty. It is occupied by many thousands of tiny bodies, the Minor Planets or asteroids, which would be mere grains of dust on the scale which we have chosen. Recently we have found many more small bodies, of asteroid size, orbiting near and beyond the orbit of Pluto – and there are grave doubts as to whether Pluto is worthy of true planetary status. I will have more to say about this in Chapter 10.

The individual motions of the bright planets have been known since very early times, and the very word "planet" means "wandering star". The ancients also noticed that the planets keep strictly to a certain region of the sky, which they named the Zodiac. The reason for this is that the paths or "orbits" of the planets lie almost in the same plane, so that when we draw a plan of the Solar System upon which a piece of flat paper, as in Fig. 4.1, we are not very far wrong. Consequently, the planets can be seen only in certain directions, and this limitation applies also to the Sun and the Moon. The Sun's apparent yearly path among the stars indicates the "ecliptic".

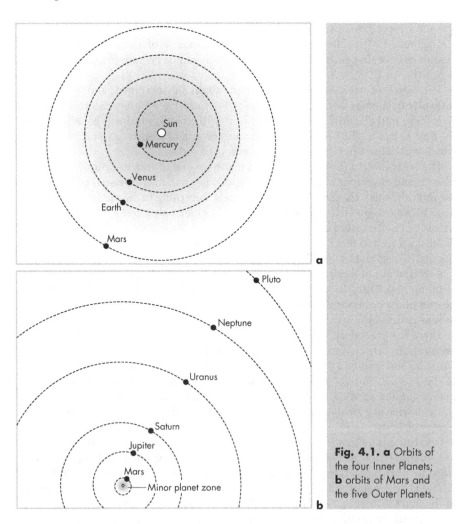

Fig. 4.1. a Orbits of the four Inner Planets; **b** orbits of Mars and the five Outer Planets.

A good way to make this clear is to imagine that we are standing in a wood, looking at low trees around us. There may be trees to all sides, but no trees will appear in the sky or beneath our feet – because the trees lie in roughly one plane, the plane of the Earth's surface.

As we know, the stars were originally looked upon as mere points of reference. The early astronomers grouped them into constellations, and there are twelve constellations in the Zodiacal band, which stretches right round the heavens. The most famous of these groups is probably Aries, the Ram. It contains no very bright stars, and is not particularly easy to identify, but in the far-off times when the Chaldaean shepherd-astronomers gazed at the skies during their night watches Aries was the constellation in which the ecliptic cut the "celestial equator", the projection of the Earth's equator upon the celestial sphere. Actually, the point of intersection, or "First Point of Aries"' has moved since then, because of the wandering of the polar point, and has now passed into the neighbouring constellation of the Fishes; but we still keep to the old term.

Since the planets are never far from the ecliptic, they are easy to recognize. In any case, Mars (when at its brightest) and Jupiter are so distinctive that they cannot possibly be confused with stars, while Mercury and Venus, which are closer to the Sun than we are, have their own way of behaving. Only Saturn, and Mars when at its faintest, cannot be identified at the most casual glance.

The first astronomer to give a proper description of the way in which the planets move was Johann Kepler. Between 1609 and 1619 he published his three famous Laws of Motion, which are interesting enough to describe in slightly more detail. They are as follows:

Law 1. The planets move in ellipses, with the Sun at one focus.

Law 2. The radius vector (the line joining the centre of the planet to the centre of the Sun) sweeps out equal areas of space in equal times.

Law 3. The square of the sidereal period is proportional to the cube of the planet's mean distance from the Sun.

These may seem rather complex, but really they are quite simple. Law 1 requires no explaining; the only point to bear in mind is that although the orbits of the planets are ellipses, they are of slight eccentricity, and do not depart much from the circular form. It is the other two Laws which sometimes cause beginners to wrinkle their brows.

Law No. 2 is explained in Fig. 4.2. The figure is not to scale, and the orbit of our supposed planet P is much more eccentric than is actually the case with any major planet in the Solar System, but one has to make the diagram inaccurate in order to make it clear! S is the Sun; P, P1, P2 and P3 stand for the planet in various positions in its orbit round the sun.

Assume that the planet moves from P to P1 in the same time that it takes to go from P2 to P3. Then the shaded area of PSP1 must be equal in area to the dotted area of P2SP3. Since the dotted area is "longer and thinner", it is clear that the planet is moving at its quickest when closest to the Sun.

This fact is vitally important. It can be summoned up by the simple rule "The nearer, the faster". The Law does not mean only that a planet moving in an elliptical orbit must travel at a varying speed; it means also that a planet when close to

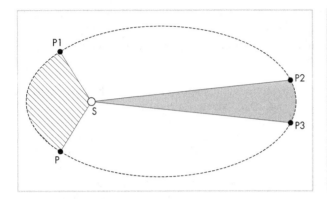

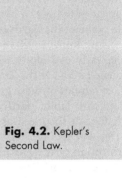

Fig. 4.2. Kepler's
Second Law.

the Sun must move faster than when it is more distant. This is borne out by direct measurement. Mercury, for instance, has an orbit which is definitely eccentric, so that at its closest to the Sun ("perihelion") it is only $28\frac{1}{2}$ million miles away, as compared with $43\frac{1}{2}$ million miles at its farthest point ("aphelion"). The orbital speed varies from $36\frac{1}{2}$ miles per second at perihelion to only 24 at aphelion. The Earth, at the greater distance of 93 million miles, is a comparative sluggard, and has an average rate of a mere $18\frac{1}{2}$ miles per second.

The Third Law leads to some equally important conclusions. The "sidereal period" of a planet, the period taken to complete one revolution round the Sun – the planet's "year" – is linked with the actual distance from the Sun, and if we know the one we can find the other.

The Earth's sidereal period is $365\frac{1}{4}$ days. By studying the way in which the other planets seem to move, we can find out their respective periods, which range from 88 days for Mercury to slightly less than 248 terrestrial years in the case of Pluto. Once this has been done, we can draw up a complete model of the Solar System in terms of the "astronomical unit", the distance between the Earth and the Sun.

To turn these relative distances into miles, all we need is one precise measurement. If, for example, we could obtain an accurate value the distance of Venus, the length of the astronomical unit could be calculated. This was done, from 1961, by radar – the general principles being to "bounce" an energy pulse off Venus, time the delay before the "echo" returns and then calculate the distance travelled, remembering that a radar pulse, like visible light, moves at 186,000 miles per second. We now know the length of the astronomical unit very seriously; in round figures it is 92,957,000 miles (149,600,000 kilometres).

The Moon, which revolves round the Earth,* is of special interest to us. Everyone is familiar with its monthly phases, from new to full and back again to new, but not everyone is sure how they are caused. Some people still believe that they are due to the shadow of the Earth, but the true explanation is far simpler.

---

*Actually, the Earth and Moon revolve round their common centre of gravity; but as this point lies within the terrestrial globe, the plain statement that "the Moon revolves round the Earth" is good enough for most purposes.

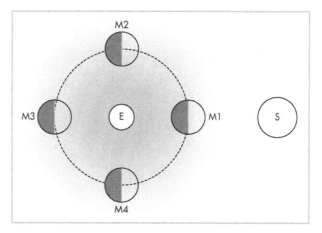

Fig. 4.3. Phases of the Moon. S, Sun; E, Earth; M1 to M4, the Moon in four different positions in its orbit. Not to scale.

The Moon is a dark body, shining only by reflected sunlight. As the Sun can light up only one half of the lunar globe at a time, the other half must be non-luminous, and therefore invisible. In Fig. 4.3, the Moon is shown in four positions in its monthly journey – M1 to M4. At M1, the dark side is turned towards us; since this does not shine, the Moon is invisible, or new. As the Moon moves on towards M2, a little of the day hemisphere starts to turn in our direction, and we see the familiar crescent shape; by the time M2 is reached, half the sunlit side is presented, and the Moon is at half phase. (Rather confusingly, this is termed First Quarter – because the Moon has completed roughly one quarter of its orbit from new to new.) Between M2 and M3 the appearance is "gibbous", between half and full, and by the time M3 is reached the Moon shows us the whole of its day hemisphere. After Full, the phase wanes once more, to half-moon at M4 (Last Quarter) and then crescent, until M1 is reached at the next new moon.

Clearly, the Earth's shadow has nothing to do with these phases. It is true that when the Moon is full (M3) and the three bodies are perfectly lined up, the shadow of our globe does fall across the Moon, causing a lunar eclipse; but eclipses do not occur every month, because the Moon's orbit is somewhat tilted with respect to ours.

The lunar phases must have been known since the dawn of history, but it was not until the invention of the telescope that Venus and Mercury were found to behave in a similar way. The phases of Venus, first detected by Galileo, are explained by Fig. 4.4. E represents the Earth, which is assumed to be stationary (really, of course, it is moving round the Sun all the time, but this makes no difference to the illustration); S the Sun, and V1 to V4 Venus in four different positions. Since Venus is closer to the Sun than we are, and moves more quickly, it completes one circuit in only 224.7 terrestrial days.

At V1 the Earth, Venus and the Sun are in a straight line, with Venus in the middle. The night side is then turned towards us, and Venus is new, so that it cannot be seen at all. This position is known as "inferior conjunction". Occasionally the alignment is perfect and Venus can be seen as a black spot against the solar disk; but since Venus too has a tilted orbit, these "transits" are rare. The next will not occur until the year 2012.

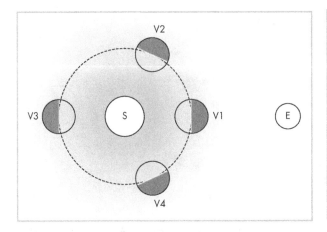

Fig. 4.4. Phases of Venus. S, Sun; E, Earth; V1 to V4, Venus in four different positions in its orbit. Not to scale.

As Venus moves on towards V2, we start to see the sunlit side. The planet appears in the morning sky as a slender crescent, becoming brighter and brighter as it draws away from the line of sight with the Sun. At V2 the three bodies form a right-angled triangle, so that Venus appears as a half disk. It then rises some hours before the Sun, and is a splendid object in the east before dawn. The technical term for this is "Western" or Morning Elongation.

As it travels towards V3, Venus changes from a half into a gibbous disk, and draws back towards the direction of the Sun so that it grows less conspicuous. By the time it has reached V3, it has ceased to be visible except during the hours of daylight. It is then at "superior conjunction", and since it lies almost behind the Sun it is not easy to find even with a telescope.

After passing superior conjunction, Venus makes its appearance low down in the evening sky, shrinking gradually to a half as its angular distance from the Sun grows. It reaches eastern elongation at V4, and is then at half-phase once more, after which it narrows to a crescent as it returns to inferior conjunction at V1.

The "synodic period" of Venus, the interval between one inferior conjunction and the next, is 584 days, though this may vary by as much as four days either way. The interval between its appearance at V4 and that at V2 is about 144 days, while 440 days are needed for the much longer interval between the appearance at V2 and that at V4.

Venus is of course at its closest to the Earth at inferior conjunction. The distance is then reduced to about 24 million miles, about a hundred times as great as that of the Moon; but as the dark side is then almost wholly presented, we cannot see the planet at all. When the disk is almost fully illuminated, Venus is a long way away. It is in fact a most infuriating object to observe.

Mercury behaves in the same manner as Venus; but since it is smaller, as well as being closer to the Sun, it is much less easy to study. It is never conspicuous to the naked eye, and only at favourable elongations can it be seen glittering near the horizon like a star. This is interesting, in view of the fact that many people believe that planets cannot twinkle. It is true that a planet, which shows a definite disk, twinkles much less than a star, which appears only as a minute point of light; but

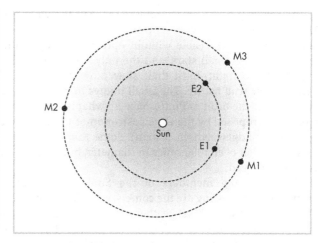

Fig. 4.5. Oppositions of Mars. E1 and M1: positions of Earth and Mars at the opposition of August 2003. E2 and M3; positions at the opposition of November 2005. There was no opposition in 2004.

when a planet is low down, and thus shining through a dense layer of atmosphere, it may twinkle violently. This is particularly so in the case of Mercury.

The remaining planets lie beyond the Earth in the Solar System, and cannot appear as halves or crescents. Mars is shown in the diagram (Fig. 4.5), and is typical of all the rest.

Let us start with the Earth at E1 and Mars at M1. The Sun, the Earth and Mars are lined up, with the Earth in the middle; Mars is therefore directly opposite the Sun, and is at "opposition". One year later, the Earth will have completed one revolution, and will have arrived back at E1; but Mars moving more slowly in a larger orbit, will not have had time to get back to M1. It will have travelled only as far as M2, and will lie on the far side of the Sun, badly placed for observation. The Earth has to catch it up, with Mars moving onwards all the time, and on an average 780 days elapse before the three bodies are lined up again. The 780-day interval between successive oppositions is therefore the synodic period of Mars.

The giant planets are much more remote, and move so much more closely that the Earth takes less time to catch them up. Jupiter's synodic period is 399 days, while in the case of far-off, sluggish Pluto the period is $366\frac{3}{4}$ days. After having completed one circuit of the Sun, the Earth has to travel on for only an extra day and a half before it catches up with Pluto.

Each of the nine major planets has its own characteristics. The members of the inner group are small, solid bodies; all have appreciable atmospheres, with the exception of Mercury; and there is still a faint probability that low forms of life flourish on Mars. The giants, however, are built up on a very different pattern. When we look at Jupiter or Saturn, what we see is not a solid, rocky globe, but the outer layer of a deep "atmosphere" made up of poisonous gases. Pluto presents problems of its own, but since it is so faint and so far away it is not not of much interest to the amateur.

Most of the planets have satellites. The Earth, of course; one – the Moon; Mars has two, both very small, Pluto has one, but may be a special case; the four giant planets have many satellites each, though only a few within the range of small or moderate telescopes. Only Mercury and Venus seem to be solitary travellers in space.

The minor planets, or asteroids, swarm in the wide gap between the orbits of Mars and Jupiter. All are dwarf worldlets, less than 700 miles in diameter, and only one (Vesta) is ever visible without optical aid. Even with a telescope, they look remarkably like small stars, and the only way to identify them is to watch them from night to night, until their slow movement across the starry background betrays their true nature. The small bodies moving in the far reaches of the Solar System, near the orbit of Pluto, make up what is known as the Kuiper Belt (its existence was proposed by the Dutch astronomer Gerard Kuiper – and slightly earlier by Kenneth Edgeworth, in Britain). Some Kuiper Belt objects are large, and one, Quaoar, is well over 1000 miles in diameter, but they are so far away that they are very faint indeed.

The remaining members of the Sun's family are much less substantial. Particularly interesting are the comets, which have been termed the stray members of the Solar System. Most of them move in elliptical orbits, but their orbits are much more eccentric than those of the planets. Fig. 4.6 shows the path of a periodical comet (Halley's) as compared with the orbit of Saturn. Nor is a comet a solid body; it is made up of a swarm of particles contained in an envelope of very thin gas. A famous astronomer once called comets "airy nothings", and though they are not "airy" in the usual sense of the word they are certainly flimsy. Comers may be of immense size, but they are of negligible mass, and they are of course completely harmless, even though they still strike terror into the hearts of some of the Earth's unsophisticated peoples.

The ghostlike nature of a comet means that it can be seen only when it is fairly close to the Earth and to the Sun. Halley's Comet – named after Edmond Halley, Flamsteed's successor at Greenwich, who was the first to realize that it revolved round the Sun – has a period of 76 years. It last came to perihelion in 1986, and will be back once more in 2061. We know where it is, but at the moment we cannot see it. Other comets have much shorter periods – only 3.3 years in the Encke's Comet, which with modern telescopes can be followed all round its orbit (even at its furthest from the Sun, it is still much closer-in than the orbit of Jupiter). There are also comets which have periods of many centuries, so that we cannot predict them; the lovely Comet Hale-Bopp, which graced our skies for months in 1997, will not return for well over 2000 years.

Meteors, or shooting-stars, are also members of the Solar System. The name is misleading, since they are not stars at all. They are small pieces of matter travelling

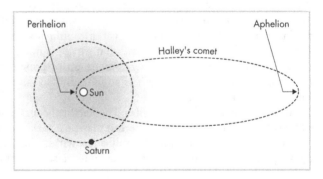

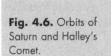

Fig. 4.6. Orbits of Saturn and Halley's Comet.

round the Sun in elliptical orbits, and in the ordinary way they are too faint to be seen. Sometimes, however, a meteor may come close to the Earth, and if it is moving at the right speed in the right direction it will naturally encounter the Earth's mantle of air. It will then plunge into the upper atmosphere, and will rub against the air-particles, setting up friction; first it will become warm, then hot, and then it will burst into flame, usually burning itself completely away in a matter of seconds and finishing its earthward journey in the form of fine dust (Fig. 4.7).

It is easy to prove that air sets up resistance. If you cup your hand and swing it abruptly, you can feel the pressure; a stick hisses through the air if swished, and the friction against the air causes a certain amount of warmth. Small wonder that a meteor, travelling at a tremendous speed, will become violently heated. Above a height of 120 miles or so, the air is of course too thin to cause appreciable resistance.

A larger body may survive the complete drop to the ground, and land more or less intact, sometimes making a crater; it is then termed a meteorite. But a meteorite has no connection with a meteor; the two classes of objects are entirely different. Shooting star meteors are cometary débris, whereas meteorites come from the asteroid zone. Some impact craters are spectacular, and there is always the chance that the Earth may be hit by a meteorite massive enough to do global damage.

Fig. 4.7. Meteors from the Leonid shower. The meteors are the short streaks crossing the picture against the background of stars. (Photograph by Michael Maunder.)

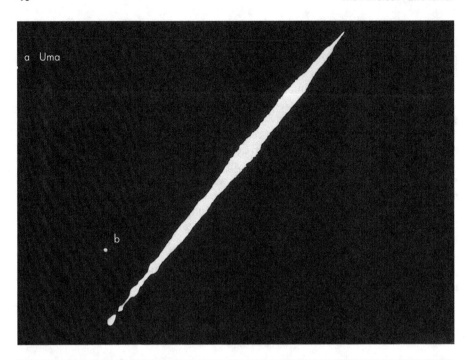

Fig. 4.8. Fireball 8/9th November 1991. John R Fletcher, FRAS Mount Tuffley Observatory, UK.

Such is the Solar System. It contains bodies of all kinds, from the vast, intensely luminous Sun down to tiny particles of interplanetary dust, and even though it may be unimportant in the universe as a whole it is of supreme importance to ourselves.

Obviously, our knowledge of the bodies of the Solar System has increased beyond all recognition since 1957, when the flight of Russia's Sputnik ushered in the Age of Space. Some of our cherished ideas proved to be right. Others were completely wrong. Men have been to the Moon, landings have been made on Venus and Mars; of the other planets only Pluto has not been surveyed from close range. Yet amateur observers can still carry out valuable work, and in any case there is endless enjoyment to be found in turning a telescope toward the craters of the Moon, the icy caps of Mars, the cloud belts of Jupiter and the glorious ring-system of Saturn.

Chapter 5

The Sun

Studying the Sun calls for methods different from those used in any other branch of astronomy. In other cases, the main problem is to collect as much light as possible; with solar observation there is plenty of light available, but it is highly dangerous to look directly at the Sun's disk using a telescope, as the eye of the observer is certain to be damaged.

The Sun's diameter is 865,000 miles, 109 times that of the Earth. But though the solar globe could contain over a million bodies the size of our own world, it does not contain the mass of a million Earths. Only 332,000 Earths would be required to make one body with the mass of the Sun. This means that the Sun is less massive than one might expect from its size, and that the mean density is less than that of the Earth – in fact, only 1.4 times as great as that of water.

Of course, this is not the uniform density throughout the solar globe. Density increases with depth. Near the centre of the Sun, the material is denser than steel, even though it is still technically a gas, whereas the outermost parts of the Sun are more rarefied than the best vacuum we can produce in terrestrial laboratories.

The gravitational force that would be felt by a man standing on the surface of a globe depends upon two factors, the mass and the size. Taking the Earth's surface gravity as unity, the surface gravity of another body can be found by dividing the mass by the square of the radius. For the Sun, these figures are respectively 332,000 and 109, so that the surface gravity is 332,000 divided by 109 squared, or 28. A man who weighs 14 stone on Earth would weigh $2\frac{1}{2}$ tons if he could be taken to the surface of the Sun, so that he would not even be able to stand upright; he would be crushed by his own weight. It is interesting to recall that Sir William Herschel, the greatest astronomer of the early 19th century, believed the Sun to be inhabited. The climate would be a little torrid; the surface temperature is 5500°C.

The great size and the low density mean that the Sun cannot be a solid body like the Earth. It is in fact made up entirely of gas, though deep down inside the globe this gas is under tremendous pressure – at least a thousand million atmospheres – and behaves therefore in a decidedly un-gaslike manner judged by our normal standards.

Telescopic views of the Sun can show interesting features, such as the dark spots and the brighter patches or faculae, but for more detailed studies we have to turn to instruments based on the principle of the spectroscope.

Newton was the first to explain the breaking-down of white light into its constituent colours. What he did was to cut a small hole in the shutter of his window, so that only a narrow beam of sunlight could pass through. This beam entered a glass prism, and the resulting rainbow or spectrum was spread out on the far wall. Later, Newton improved the experiment by using a slit instead of a hole, and by putting a lens between the prism and the wall so that he could bring the colours to a sharp focus.

Newton never took his investigations much further, probably because his prisms were of poor quality glass. The next development was due to Fraunhofer, who returned to the problem in 1814, and who found that the spectrum of the Sun is crossed by numbers of dark lines of different degrees of intensity. It is now known that each of the Fraunhofer Lines is due to the effect of one definite substance, and this is the basis of all solar and stellar spectroscopic work. One substance (such as iron) may produce many characteristic lines.

It may be added that dark lines had been seen in 1802 by a British scientist, Wollaston. Wollaston did not however realize their importance, and thought that they merely marked the boundaries of the different spectrum colours, so that the main credit must go to Fraunhofer.

All matter in the universe, whether in the Earth, the Sun or the remotest star, is made up of different combinations of a small number of fundamental "elements". There are 92 familiar elements, hydrogen being the lightest and uranium the heaviest; since they form a complete series there is no chance of our having missed one. No new elements can exist, because there is no room for them in the sequence; one might as well try to fit an extra integer between 7 and 8, or a new musical note between F-sharp and G. (It is true that various extra elements have been made artificially in recent years, but these "lead on" from the end of the sequence, and probably do not occur naturally.) We can thus be certain that each Fraunhofer Line is due to an element or group of elements already known to us.

When observed with the aid of a prism or spectroscope, the bright surface of the Sun, the photosphere, gives the bright rainbow studied by Newton. Above this is a layer of incandescent vapour, extending upwards for perhaps 10,000 miles. On its own, this vapour would give not a rainbow, but a number of bright isolated spectrum lines. However, there is the bright background to be taken into account, and the result is that instead of appearing bright, the lines emitted by the upper vapour are "reversed", and seem to be dark. This "reversing layer" is the outer envelope, or chromosphere.*

The dark lines give us a key to the elements responsible for them. The spectra of the various elements have been studied in terrestrial laboratories, and the positions of the lines are known with high accuracy, so that all we have to do is to compare the laboratory lines with those visible in the solar spectrum. If a solar line corresponds to a laboratory line of sodium, we can prove that there is sodium in the Sun. In this way nearly 70 of the 92 familiar elements have already been identified.

---

*Many books differentiate between the "reversing layer" and the "chromosphere," but there is no justification for this. The whole chromosphere is a reversing layer, though all the solar elements occur in the lower part of it, so that this part gives the most complete bright-line spectrum.

We are now in a position to examine the structure of the Sun itself. Near the centre of the globe, the pressure is tremendous, while the temperature is terrifyingly high – something like 1.5 million degrees Centigrade, which is beyond our comprehension. It is here that the production of energy is going on, and the inner region has aptly been termed "the solar power-house".

The visible surface of the Sun, the photosphere, is the region where the solar gases become thin enough to be transparent. The bright rainbow spectrum originates in the photosphere, and here too we meet the curious dark patches which are known as sunspots. Any small telescope will show them (but beware of the dangers of solar observation – see below), and it is fascinating to track them as they drift slowly across the Sun's disk, and to watch their shapes change from day to day.

We have found out a great deal about the way in which sunspots behave, and we also know that they are associated with magnetic phenomena. Broadly speaking, a spot may be described as a relatively cool patch on the photosphere, so that it emits less light than the surrounding surface. "Cool", however, is here in the solar and not the terrestrial sense; the mildest part of a spot has a temperature of some 4,000°C, but the difference between this and the normal photosphere is enough to make the spot appear dark. If seen by itself, it would however glow with a brilliance much greater than that of an arc-lamp, so that it would be a grave mistake to describe a sunspot as "black".

A large spot is made up of a relatively dark central portion (umbra) and a lighter surrounding area (penumbra). Several umbrae may be contained in one mass of penumbra; sometimes the shape of the whole spot is circular, sometimes the outline is complex and irregular. Small spots may be made up entirely of umbra, while in complex groups the penumbral area is widely scattered (Fig. 5.1).

Spots may appear singly, but more often form groups. A common sight is to see two main spots, one lying to the west of the other, with numerous smaller ones nearby. In general, the following or easterly spot is the first decay and vanish. A really large group may contain dozens of separate umbrae, and sometimes the detail is so intricate that it is difficult to photograph and almost impossible to draw.

The average spot lasts for about a week before it disappears, while smaller ones may have a lifetime of less than a day. Occasionally an unusually persistent spot makes its appearance; the record for longevity seems to belong to a spot which was seen from June to December 1943, a total period of nearly 200 days. The spot was not, of course, under continuous observation for the whole of that period. Since the Sun rotates on its axis, taking rather less than a month to do so, a spot group can be seen moving slowly across the disk as it is carried from east to west. The movement is too gradual to be noticed over short periods, but the shift from one day to the next is very obvious indeed. After a time, the spot will be carried over the western limb, and will not be seen again for a fortnight, after which it will reappear in the east – if, of course it still exists.

Sunspots are essentially magnetic phenomena. The Sun's lines of magnetic force run from one magnetic pole to the other, below the visible surface; spots are produced in places where the lines break through to the top of the photosphere.

The numbers of sunspots vary in a semi-regular cycle. Maxima, during which spots are frequent, occur at intervals of about 11 years, with minima in between. During an active period there may be as many as a dozen groups visible at once,

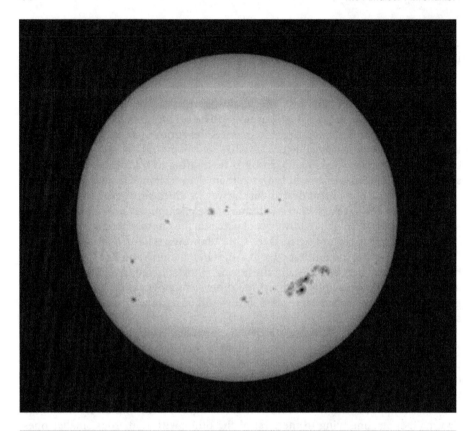

Fig. 5.1. Sunspots: 30 March 2001. This picture shows the Sun in white light. (Photograph by John Leibacher of the National Science Foundation's Global Oscillation Network Group and the National Solar Observatory, NSF, GONG and NSO, respectively.)

while near minimum the whole Sun may be spotless for weeks on end. This cycle was discovered by a German amateur, Heinrich Schwabe, who drew the Sun on every possible day between 1825 and 1843, counting the observable spots and studying their individual characteristics.

Actually, the cycle does not keep strictly to the 11-year period. The interval between successive maxima may be as short as 9 years, or as long as $13\frac{1}{2}$, so that no exact forecasts can be made; moreover, some maxima are more active than others. There was even one period, between 1645 and 1715, when there were very few spots, so that the solar cycle was suspended for reasons which are unclear. It is known as the Maunder Minimum, since it was studied from records by the English astronomer E W Maunder. The last maximum fell in 2000, so that we may expect the next one in 2011; around 2006–7 spots are likely to be in short supply and activity as a whole will be at a comparatively low ebb.

Regular observation will show that the spots do not appear to move across the Sun in straight lines, except during early June and early December. This is because

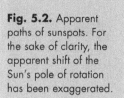

Fig. 5.2. Apparent paths of sunspots. For the sake of clarity, the apparent shift of the Sun's pole of rotation has been exaggerated.

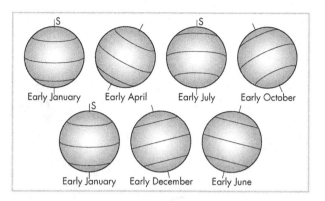

Early January　Early April　Early July　Early October

Early January　Early December　Early June

of the apparent shift in position of the Sun's axis of rotation. The position of the pole for any date can be looked up from the tables contained in a publication such as the *Handbook of the British Astronomical Association*, but the rough diagrams in Fig. 5.2 may be of help.

During the early part of a cycle, the spots tend to appear some way from the equator, but as the cycle progresses the spots invade lower and lower latitudes. As the cycle draws to its end, and its group die away, small spots of the new cycle start to appear in high latitudes once more. At minimum, therefore, there are two areas subject to spots: the equatorial, with the last spots of the dying cycle, and the higher-latitudes, with the first spots of the new cycle. This behaviour is termed Spörer's Law, since it was first announced in 1879 by the German astronomer of that name; it is extremely important to solar physicists. It should be added that spots never break out near the Sun's poles of rotation.

Spots are associated with bright irregular patches known as faculae, from the Latin word meaning "torches". Faculae appear to lie well above the photosphere, and can be regarded as luminous clouds hanging in the upper regions. They often appear in positions where a spot-group is about to break out, and they persist for some time after the group has disappeared. Consequently, the appearance of faculae on the Sun's following limb is often an indication that a spot group is coming into view from the far side.

Sunspots possess strong magnetic fields, and emissions from the active regions lead to disturbances of the compass needle, as well as displays of aurorae or Polar Lights (Fig. 5.3). Whether sunspots have any effect on our weather is debatable – though it is worth noting that the Maunder minimum coincided with a very cold period at least in Britain; during the 1680s the Thames froze every winter, and frost fairs were held on it.

Even on the unspotted areas of the Sun constant activity is going on. The photosphere is covered with granules – vast convective cells of hot gases, rising and falling rapidly, and even of the order of 600 miles across; on average each granule lasts for around 8 minutes. No part of the surface is free from them.

Very occasionally, brilliant short-lived patches may be seen over sunspots. The first of these "flares was seen in 1859 by two amateurs, Carrington and Hodgson,

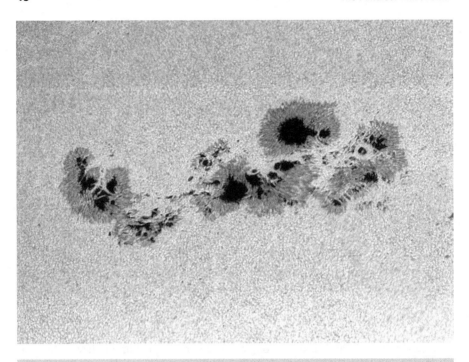

Fig. 5.3. Close-up on the group of sunspots: 30 March 2001. (Photograph by W. Livingston of NSF's GONG and NSO.)

but for many years no more recorded. Flares visible in ordinary telescopes are in fact so rare that an observer may go through his whole lifetime without seeing one, but modern instruments have shown that the solar flare is a common phenomenon.

Flares are phenomena of the chromosphere or 'colour-sphere', the region of the Sun which lies above the photospheres, and are, naturally, commonest near the times of solar maxima. They may be described as being storms in the chromosphere, of an electrical nature, the hydrogen storms being caused to glow brilliantly by electrical excitation. They spread through large areas of the chromosphere horizontally, i.e. parallel with the solar surface, with amazing rapidity, but there is very little vertical movement; they seem to be confined to the 8,000 or 10,000 miles of the chromospheric depth. They produce marked effects upon the terrestrial compass-needle, as well as helping to cause radio fade-outs and other disturbances (Fig. 5.4).

Except during a total solar eclipse, when the Moon temporarily hides the disk of the Sun, the chromosphere and the corona, which lie beyond, cannot be seen visually, because they are hidden by the intense glare of the photosphere. Instruments based upon the principle have to be used. So for the moment, let me deal with features of the Sun which can be seen with ordinary telescopes.

Observing the Sun is not the simple matter that might be imagined. Even a small telescope can concentrate so much light and heat that an incautious observer who puts his eye to the tube may be blinded. Very great care is necessary at all times; it is only too easy to make a mistake.

Fig. 5.4. Solar Prominence. (Photograph by courtesy of ESA/NASA.)

Unfortunately, it is possible to buy special dark-lensed "suncaps" which fit over an ordinary eyepiece, and can be used for direct observation. According to some textbooks, it is then safe to turn a 2- or 3-inch refractor directly towards the Sun, and observe in the usual manner. *This is emphatically not the case.* No suncap can give full protection, and in any case there is always a chance that the cap will splinter, so that the eye of the observer will be seriously injured before he has had time to realize what has happened. This warning is not mere alarmism; I know of one amateur who lost the sight of his left eye through an accident of this sort, and the risk is not worth taking, particularly when better observations can be made by indirect means.

There is another danger also. Sometimes the Sun can be seen shining through a layer of thick mist, so that it appears reassuringly dim and gentle. The temptation is then to use a telescope directly, either with or without suncap. Here again there is more than a chance that permanent damage to the eye will result; as soon as the solar radiation is focused, it becomes unsafe. There are various wedges and filters which are safe enough if properly used, but I do not recommend them unless you have expert knowledge. By far the best way to draw sunspots is to project the Sun's image on to a piece of white card (see Fig. 5.5).

Projection is an easy process, since there is plenty of light available. First turn the telescope in the direction of the Sun, "squinting" over the top of the tube and keeping a cover over the object-glass. Then rack out the focus, and remove the cap from the end of the tube. Hold a white card a few inches away from the eyepiece, and move the telescope gently (if necessary) until the image of the Sun appears, after which the disk can be brought to a sharp focus by adjusting the rack and the position of the card (Fig. 5.6). Any spots and faculae that happen to be present will be obvious at a glance. A low power is advisable – I have found that for my 3-inch

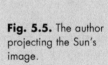

Fig. 5.5. The author projecting the Sun's image.

f/12 refractor, ×72 gives good results – though the magnification can be increased for drawings of individual spots on a larger scale.

To make the drawings conveniently standard, it is as well to draw a 6-inch circle on a card and then adjust the distance and focus until the image of the Sun exactly fills it. If the telescope used is very small, a 4-inch circle may have to suffice.

It is not easy to hold the card steady, move the telescope to follow the Sun, and draw the visible spot-groups at the same time. One would have to be a Briaraeus to

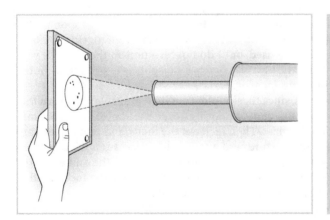

Fig. 5.6. Simple projection of sunspots, using a 3-inch refractor.

do so effectively, and the obvious solution is to fit an attachment to the telescope tube which will hold the card at the right distance from the eyepiece. Such an attachment can be built by anyone who is reasonably skilful with his hands, and there is nothing in the least difficult about it. The main thing to avoid is upsetting the balance of the telescope tube.

When the drawing has been finished, the following details should be added: date, time (G.M.T.: never Summer Time), observer's name, aperture of telescope, and magnification. If any of this information is omitted, the drawing promptly loses most or all of its value.

In general, refractors are to be preferred to reflectors for solar work, and the ideal aperture is from 4 to 5 inches. A 6-inch is larger than is necessary, and extra care must be taken. In the case of a reflector, the mirror should be left unsilvered, which naturally makes the instrument almost useless for any other kind of work. During the many years that I have owned my 12½-inch reflector, I have never turned it towards the Sun, and nor shall I ever do so.

If observations are made daily, it is interesting to work out what is called the Zürich number, which gives the degree of activity at that particular time. Each separate spot-group gives 10 points, and each separate umbra 1; sometimes the Zürich number is very high, while for a spotless disk it is of course zero. Watch too for the Wilson effect. If a sunspot has a depressed umbra, then the "preceding" penumbra will be foreshortened, and will appear narrow as the spot comes over the limb, while when the spot has crossed the disk and is approaching the opposite limb the "front" penumbra will appear to be the broader. In fact, with a circular sunspot with a depressed umbra, the penumbra closest to the Sun's central meridian will always seem to be relatively narrow. Not all spots show the Wilson effect; no two groups are identical.

While it would be idle to pretend that the observer who contents himself with drawing sunspots with the aid of a small refractor has much chance of making a valuable discovery, particularly since daily disk photographs are taken at solar observatories, the time spent will not be wasted. Much will be learned; it is fascinating to watch the spots and faculae as they drift, change and finally die away. Yet we must never forget that we are unworthy to take liberties with the ruler of the Solar System. A cat may look directly at a king, but no telescopic worker must ever look directly at the Sun.

To take matters further, you will need more specialized equipment. The telescope shown here is fitted with filters which cut out all the light except that of hydrogen, and it will show flares, active areas, and prominences. It does mean an outlay of well over £1000, but the cost is non-recurring, and there is always plenty to see, and it does mean that the solar enthusiast can observe in comfort and warmth rather than tolerating the cold and darkness of a bitter winter night!

Chapter 6

The Moon

The Moon is much the closest natural body in the sky. On average, it is a mere 239,000 miles away from us; and although it is smaller than the Earth, with a diameter of only 2,160 miles, it dominates the scene during the hours of darkness. It is hardly surprising that our ancestors worshipped it as a god.

Definite markings can be seen with the naked eye, and any telescope or pair of binoculars will show a vast amount of detail. There are mountains, valleys and craters; the sight of a lunar landscape is something never to be forgotten, and the Moon will always be the favourite object for amateur observation. Moreover, amateurs have carried out very useful work in lunar charting. It is probably true to say that before the start of the Space Age, the best of all Moon maps were of amateur construction.

The situation today is very different. The Moon is no longer inaccessible; it has been reached, and many of its outstanding problems have been cleared up, though many more remain to be solved. Quite apart from the manned landings, there have been automatic probes which have flown round and round the Moon, securing photographs of amazing detail and quality, so that by now we have extremely accurate charts of the entire surface.

In astronomy, as in everything else, honesty is the best policy, and it is best to admit immediately that in most ways, though not all, the amateur lunar observer has completed his task. There is now no scientific value in, say, making a chart of a limb area with the aid of a 6-inch or even a 12-inch telescope. It is still worth doing, for the pleasure that it gives the observer; but modern lunar mapping is carried out from beyond the Earth. Only in a few restricted fields is original lunar research still within the scope of the amateur. The last thing I want to do is to be discouraging – and, as a personal aside, I have always been more concerned with observation of the Moon than with any other branch of astronomy. But there is no point in not facing facts (Fig. 6.1).

Before going into the story of how the Moon has been explored, it may be best to give a brief description of the lunar world itself. It is usually called the Earth's satellite, but in my view, at least, this is misleading; it is too large to be a satellite. Remember, its diameter is more than one-quarter of that of the Earth (Fig. 6.2), and there now seems no doubt that it is exceptional. Four planetary satellites – three in Jupiter's family, one in Saturn's – are larger than the Moon, but all three

51

Fig. 6.1. The full moon, my photo, 1998.

orbit giant primaries, and it may be sensible to regard the Earth-Moon system as a double planet.

There is still some uncertainty about the origin of the Moon. The old theory, that it used to be part of the Earth and was simply thrown off into space, has been abandoned on mathematical grounds. Possibly the Earth and the Moon were formed at about the same time in the same region of space; certainly they are of the same age

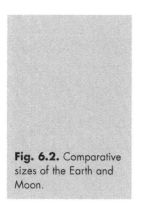

Fig. 6.2. Comparative sizes of the Earth and Moon.

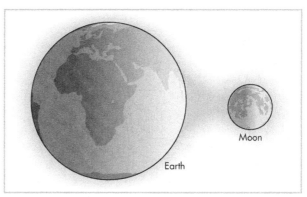

Moon

Earth

(4.6 thousand million years). Today there is strong support for the 'massive impact' theory, according to which there was a collision between the original Earth and a body about the size of Mars; the cores of the Earth and the impactor merged, and mantle débris ejected during the collision accreted to form the Moon. This does at least explain why the overall density of the Moon is much less than that of the Earth.

It is not strictly true to say that the Moon moves round the Earth; more accurately, both bodies move round the barycentre, or centre of gravity of the system. However, the barycentre lies within the Earth's globe, since the Earth is so much more massive than the Moon; the ratio is 81 to 1. The Moon is the main force in the raising of the tides. In this respect it is much more effective than the Sun, simply because it is so much closer to us.

The Moon's low mass means that it has also a low escape velocity: $1\frac{1}{2}$ miles per second, instead of 7 miles per second as for Earth. In its younger period it may have had an atmosphere, but the weak gravity has meant that virtually all this atmosphere has leaked away into space. A trace remains, but the density is very low, and the weight of the entire lunar atmosphere cannot be more than 30 tons. Go to the Moon, as the Apollo astronauts have done, and the sky will be black all the time; shield your eyes from the glare of the rocks, dark-adapt, and you will be able to see stars in the daytime.

The Moon is a slow spinner. Its rotation period is 27.3 days, which is the same as the time taken for the Moon to go once round the Earth (or, properly speaking, around the barycentre). The effect of this synchronous or "captured" rotation is that the Moon keeps the same face turned towards us all the time. This sometimes causes confusion, but a simple experiment will show what is meant. Place a chair in the middle of the room, to represent the Earth, and assume that your head is the Moon. Your face stands for the familiar hemisphere, while the back of your neck represents the "back" of the Moon (Fig. 6.3). Now walk round the chair, keeping your nose turned toward it all the time. When you have completed one circuit, you will have turned round once; your "sidereal period" will have been equal to your "axial rotation", and anyone sitting on the chair will not have seen your back hair at all. This is how the Moon behaves. Just as the seated observer failed to see the back of your neck, so the terrestrial observer can never see the far side of the Moon.

Fig. 6.3. A simple demonstration of the movement of the Moon round the Earth.

However, there is one modification. Though the Moon spins on its axis at a constant speed, it has a somewhat eccentric orbit, and this means that its velocity varies. When at its closest to the Earth, it moves quicker than when more distant. The axial spin and the position in orbit become periodically out of step, and the Moon seems to rock slowly to and fro, allowing us to view first one limb and then the other. On some nights, the grey plain of the Mare Crisium will appear to be almost touching the limb, while on others it will be well clear, and more details will come into view. There is also a rocking in a north-south direction, since the Moon's orbit is tilted, and we can peer for some distance beyond alternate poles. These rocking motions or "librations" mean that from Earth we can examine a total of 59 per cent of the total surface, though, of course, never more than 50 per cent at any one time. The remaining 41 per cent is always out of view. Until the first circum-lunar rocket sent back photographs, in 1959, we had no positive information about "the other side of the Moon".

There is one more point to be borne in mind. The Moon always keeps more or less the same face turned to the Earth, but it does not keep the same face turned towards the Sun, so that day and night conditions are the same everywhere on the Moon's surface; it is quite wrong to suggest that there is a part of the Moon which is always dark. The only real difference is that from the far side, the Earth will never be seen, so that the nights will be blacker due to the absence of Earthlight.

Of the lunar features, the most immediately obvious are the dark grey plains which are always known as seas (Latin, *maria*). They were first charted teles-copically by Harriott in 1609, and shortly afterwards a map was produced by Galileo. It was natural to suppose that the plains were water-filled, even though Galileo himself apparently had doubts; and the romantic names are still used, even though we have long since found that there is no water on the Moon. One cannot have liquid water in the absence of atmosphere; and analyses of the Apollo samples seem to prove that the lunar seas were never water-filled (Appendix 10, Fig. A10.2).

Many of the seas, such as the vast Mare Imbrium* (Plate XVI) are roughly circular in form. Their boundaries are raised, and form mountain chains, some of which
are high by our standards. For instance, the Lunar Apennines, which form part of the
border of the Mare Imbrium, have peaks rising to 15,000 feet. Other parts of the
border are formed by the Alps, cut through by a magnificent valley, and the Caucasus
Mountains, which separate the Mare Imbrium from the neighbouring plain. Earth-
type mountain chains are rare; but isolated peaks and clusters of hills are almost
innumerable. The very highest mountains on the Moon exceed 25,000 feet. Their
altitudes are measured by the shadows which they cast, though the situation is
complicated by the fact that on the waterless Moon there is no sea-level to serve as
a standard for reference. Formerly, amateurs carried out valuable work in this direction, though we must concede that space-research methods have now taken over.

The whole scene is dominated by the walled circular formations which we usually
call craters. No part of the Moon is free from them. They cluster in the bright uplands,
and are also to be found on the maria, on the slopes of mountains and even on the
crests of the peaks. They break into each other and deform each other; some have
massive, terraced walls and high central mountain structures, while others are low-
walled and ruined, so that they come into the category of "ghosts". The very largest of
them exceed 150m miles in diameter. The smallest are too minute to be seen at all
from Earth. In general, they are named after famous personalities of the past.
Features on the far side of the Moon, invisible from Earth, have also been named. A
full list was approved by the International Astronomical Union during its meeting in
England in the summer of 1970; a few extra names have been added since.

Though some of the craters are deep, with walls rising to well over 10,000 feet
above the floors, they are not in the least like steep-sided mine-shafts. A typical
large crater has a mountain rampart which rises to only a moderate height above
the outer country, but much higher over the sunken interior. The inner walls are
often terraced, so that a lunar crater has a profile more like that of a saucer than a
well. This is shown in Fig. 6.4, in which the famous 38-mile crater Eratosthenes is

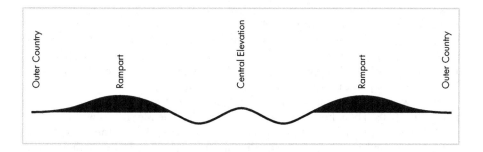

Fig. 6.4. Cross-section of the lunar crater Eratosthenes. The diagram is simplified for the
sake of clarity; the curvature of the lunar surface is neglected, and the central elevation is
shown as a simple peak instead of a complex mass.

---

*It seems best to keep to the Latin names, which are always used in astronomical literature. A full list of
these, together with their English equivalents, is given Appendix 10.

drawn. Like many other walled formations, Eratosthenes has a central elevation which may look inconspicuous, but which is really lofty. With some craters, the central structure is a single peak; inside other formations we see clusters of hills, while central craterlets are also frequent. On the other hand, there are also formations whose floors are flatter, and which have no central structures. The dark-floored, 60-mile Plato is an example of this (Appendix 10, Fig. A10.2). Like most of its kind, Plato is circular; but as seen from Earth it seems oval, because of fore-shortening. Photographs taken from space-craft show it in its true shape.

It used to be thought that the craters were of internal origin, basically similar to volcanic calderae, but we now know they are impact formations, and we have a good idea of the past history of the Moon. Between 4400 and 4000 million years ago came the Great Bombardment, when meteorites rained down to produce great basins. The last major basin, now filled by the Mare Imbrium, dates back 3850 million years, and as the Great Bombardment ended there was widespread vulcanism; magma poured from below the crust and flooded the basins as well as the floors of craters such as Plato. The lava-flows ended, and since then the Moon has seen little activity apart from the formation of occasional impact craters such as Tycho and Copernicus. There are definite differences between the Earth-turned and the far side of the Moon; the crust is thicker on the far side, and some of the basins are unflooded. One huge formation, Tsiolkovskii, seems to be a hybrid; it has a flooded, Mare-type floor, but high walls and an imposing central peak. It is a pity that it can never be seen from Earth.

Of the minor features of the surface, note in particular the clefts of rills*, which look like surface cracks and which extend in some areas for over 100 miles. A particularly good example is the long rill associated with the crater Ariadaeus; close behind it is another so-called cleft, that of Hyginus, which is made up partly of small craterlets. Look, too, at the magnificent cleft inside the majestic ringed plain Petavius, running from near the central mountain across the wall. Over the whole of the Earth-turned face, there are many clefts and cleft-systems within range of modest telescopes.

Quite different are the domes, which may be likened to gentle swellings in the crust; many of them have summit craterlets, and some riddled with fissures. And, of course, there are the bright rays which spread out from some of the prominent craters. Tycho, in the southern uplands, is the centre of a ray-system which extends in all directions for hundreds of miles, while another major centre is Copernicus in the Mare Nubium. The rays are surface deposits, and best seen only under high illumination – a point to which I will return presently.

Very few professional astronomers paid much attention to the lunar surface before the 1950s. Then, however, it became clear that the Moon was within reach, and there was a prompt upsurge of professional activity. Most of the groundwork had been carried out by amateurs, who had performed nobly, and there were various maps which were of high accuracy by the pre-Space Age standards. The next step was to undertake a full photographic survey, and work was begun in various countries, notably the United States. Detailed photographic atlases appeared, superseding the older visual charts.

---

*Many people use the German spelling, *rilles*.

Then, in 1959, came the first Moon-rockets. The first three were Russian; Lunik 1, which by-passed the Moon, was the pioneer vehicle in January. It was followed by Lunik 2, which crash-landed on the surface, and then, in October, by Lunik 3, which went round the Moon and sent back the first photographs of the far side. Today, the Lunik 3 pictures look very blurred, but that they were an immense technical triumph cannot be doubted.

The next step was taken by the Americans, with their Ranger programme. The scheme was to crash a probe on to the Moon without any attempt to preserve it. Before impact, it would send back close-range television pictures. There were several failures, but success came on July 31, 1964, with Ranger VII which came down in the Mare Nubium. The photographs were excellent, as were those from the succeeding Rangers, Numbers 8 and 9. Ranger 9 was aimed at the large crater Alphonsus, in which a certain amount of volcanic activity had long been suspected.

The year 1966 was important in lunar research. First there was Luna 9, a Russian triumph. The vehicle landed on the Moon, in the Oceanus Procellarum, but did not destroy itself as the Rangers had done; it came down gently, so that after arrival it was able to go on transmitting information. The pictures it sent back were the first to be obtained direct from the surface of another world, and they showed a terrain which looked remarkably like a lava-field. The second important development of 1966 was an American success. On August 10, Orbiter 1 was launched, and put into a path around the Moon. Photographs were received from both it and its four successors, and at once all Earth-based charts were made obsolete. By the end of the programme in January 1968, when Orbiter 5 was deliberately crashed on to the surface, the main task of mapping the Moon was more or less complete.

Of course all these experiments were leading up to the first manned missions, and, as everyone knows, the landing was achieved on 21 July 1969 by Neil Armstrong and Buzz Aldrin, in the lunar module of Apollo 11. They came down in the Mare Tranquillitatis, not far from the small but well-marked crater Möltke, and were watched by television viewers all over the world. Other Apollos followed, and by the time that the programme ended, with the return to Earth of Apollo 17 on 19 December 1972, twelve men had walked on the Moon.

Apollo 13, of 1970, will also be remembered – but for a different reason. Everything went wrong. During the outward journey there was a violent explosion, which robbed the spacecraft of its main power sources. The lunar landing, scheduled for the upland area of Fra Mauro (Appendix 10, Fig. A10.2) was abandoned, and it was only by a combination of brilliant improvisation, courage and skill that the astronauts returned unscathed.

The last three Apollo missions were undoubtedly the most valuable scientifically. On each trip the astronauts took their own transport, and the "Moon Rovers" functioned admirably. The Apollo programme ended in 1972 with No 17, and it was then that Dr Harrison Schmitt, a geologist who had qualified as an astronaut, made the discovery of "orange soil" near a small crater which had been nicknamed Shorty. Yet the orange colour did not indicate recent vulcanism, as had been thought at first; it was due to large numbers of small orange-coloured pieces of glass. Moreover, it was very old. Just why it was so localized remains a mystery.

Quite clearly the Apollo missions have increased our knowledge of the Moon beyond all recognition. Neither have the Russians been idle; in 1970 they sent up the first automatic probe which collected lunar samples and returned home, and later in the same year they dispatched Luna 17, which was even more spectacular. From it crawled Lunokhod 1, which looked like across between a surrealistic saucepan and an ancient steam-car, but which explored the whole area with amazing efficiency. Lunokhod 2 was equally successful – yet the Russian manned landing plan was a complete failure, and after the Apollo 11 success the USSR gave up.

There have been two major American missions since Apollo: Clementine (launched 25 January 1994) and Prospector (6 January 1998). Clementine – named after the character in the old song who is "lost and gone forever" – was essentially a mapping probe, and orbited the Moon until 19 February, when it was dispatched on an ultimately unsuccessful rendezvous with a small asteroid, Geographos. Clementine's pictures of the Moon were superb, but there was one surprise: it was claimed that some of the results indicated the presence of ice on the floors of some of the polar craters, which are always in shadow and which are always intensely cold. To me this always seemed widely improbable; all the rocks brought back by the Apollos and Russian probes were completely devoid of any hydrated material – and how could the ice have been deposited there? An impact by an icy comet would generate too much heat. In any case, what Clementine detected was not actual ice, but hydrogen, which could come from the solar wind. Similar claims were made for Prospector, which also was an orbiter. On 31 July 1999 Prospector was deliberately crashed into a polar crater and the investigators hoped that the debris thrown up would show traces of water. The results were negative, but at least the episode did NASA's funds a great deal of good!

By now the idea of a lunar base is by no means far-fetched, and should become reality within the next decade or two. It will be international; so far the missions have been either American or Russian, apart from one small Japanese probe (Hagomoro), but all the space programmes are now, mercifully, combined.* So is there anything left for the amateur lunar observer to do? The answer is "yes", and in particular work is needed upon time-dependent phenomena. The Moon is not active world, but tiny changes do occur, and this is where the amateur comes into his own.

However, great care is needed, and the essential first step is to learn one's way around. It is hopeless to start any systematic programme until all the main features, and many of the minor ones, can be recognized on sight. Crater identification is not so difficult as it may seem, but various basic points have to be borne in mind. When a crater is near the terminator, or boundary between the daylight and night hemispheres of the Moon (Fig. 6.5), it will have shadow inside it, and will be strikingly conspicuous; under high illumination, the shadows will shrink, and the crater may become hard to find. Toward full moon the shadows almost disappear, and even large craters are difficult to locate unless they have particularly bright walls (such as Aristarchus), particularly dark floors (such as Plato) or ray systems (such as Tycho). It is therefore wrong to suppose that full moon is the best time to start observing; on the contrary, it is the worst except for

---

*It is quite incredible to realize that there are some people who genuinely believe that the lunar landings never happened and that NASA faked the entire Apollo programme in an elaborate film set. The idea came from a science-fiction film, *Capricorn One*, in which NASA was sporting enough to collaborate, but nobody could ever have dreamed that the idea would be taken seriously. All that can really be said about these curious folk is that if ignorance is bliss, they must be very happy!

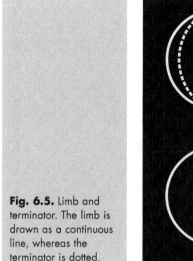
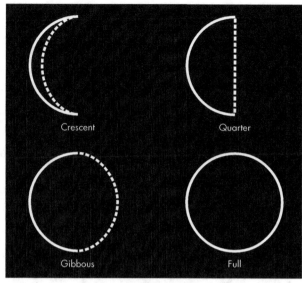

Fig. 6.5. Limb and terminator. The limb is drawn as a continuous line, whereas the terminator is dotted.

special investigations, particularly as the bright rays dominate the scene completely. The most spectacular views are obtained during the crescent, half and moderately gibbous stages.

I can cite a personal experience here. I first looked at the Moon through a telescope when I was a boy of eight, and since I knew no better I decided to make a start on the night of the full moon. I looked up the position of the 92-mile crater Ptolemaeus, arranged my newly acquired 3-inch refractor, and tried to find my way about. Naturally, I failed to find Ptolemaeus. When I looked again at the time of half-moon, the crater was partly filled with shadow, and I could identify it at first glance. The method I then adopted – and which I still recommend – was to obtain an outline chart of the Moon and then set out to make at least two sketches of every named feature. The procedure takes a long time, because a normal crater can be well sketched only when there is some shadow inside it and one has to make the most of one's opportunities. By the time I had finished it had taken me more than a year. The sketches themselves were useless, as I knew they would be; but at least I had learned how to tell one crater from another. The map I used was Elger's, published in 1896. Since then I have drawn a slightly larger outline map, though it too makes no pretence of being anything more than a guide (Figs 6.6 and 6.7).

It is a great mistake to make a drawing too small, or to attempt too large an area at one time. Probably about 20 miles to the inch is a good scale. "Finished" drawings look attractive, but an observer with no artistic gifts, such as myself, may be wiser to keep to line drawings. Accuracy is always the main objective. Always remember that a crater alters in appearance according to illumination, so that it is necessary to identify it under all possible conditions of lighting (Fig. 6.8).

Incidentally, some decisions have caused a good deal of confusion. "East" and "west" have always been standardized so that, for instance, Mare Crisium is near the west limb. The American space authorities have reversed this, making east west and west east; they also put north at the top. In the present book I have followed

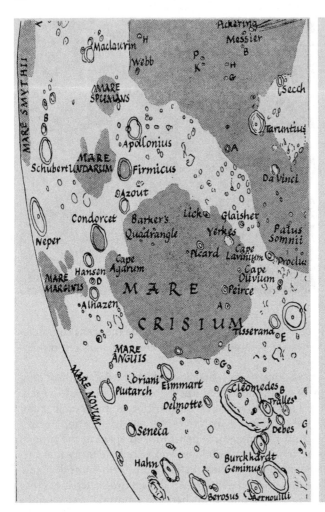

Fig. 6.6. Section from the author's 2-foot map of the Moon.

the American east-west practice, because it has been accepted by the Lunar Commission of the International Astronomical Union. (Only a few rebels, such as myself, voted against it!) However, I have kept south at the top, as has always been customary (Fig. 6.9).

It used to be officially laid down that the Moon is entirely changeless. Admittedly there was one alleged case of alteration – in the formation Linné, on the Mare Serenitatis, which was drawn as a crater by all observers before 1843, and since 1866 has been a small craterlet surrounded by a white patch – but the evidence was most unsatisfactory, and the other reported instances were even more uncertain. It seems that large-scale alterations on the Moon belong to the remote past. Yet in recent years there have been observations of a different kind, and this brings me on to the whole question of TLP or Transient Lunar Phenomena – a term which I introduced long ago and which has now come into general use.

All serious and persistent lunar observers know about the glows, and obscurations seen from time to time in localized areas. They are not easy to detect; few of

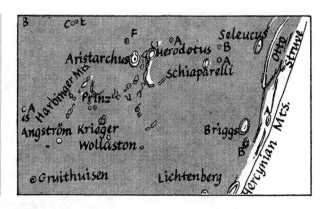

Fig. 6.7. The Aristarchus area, from the author's 2-foot lunar map.

them last for long, and it is only too easy to be deceived. There was initially a great deal of scepticism about them, mainly because in the pre-Space Age period many (not all) of the TLP reports came from amateur observers, who knew the Moon a great deal better than most professionals. Then, on 3 November 1958, the Russian astronomer N. A. Kozyrev, using the 50-inch reflector of the Crimean Astrophysical Observatory, saw a red glow inside the walled plain Alphonsus, and managed to obtain a spectrum. At once the professionals began to take TLP seriously, and in 1963 NASA produced a comprehensive catalogue of all TLP reports (the work of Barbara Middlehurst, Jaylee Burley, Barbara Welther and myself). There were hundreds of reports, and while many of them were certainly spurious, due to honest mistakes in observation, others were not.

What are they? They are not violent eruptions; major vulcanism on the moon ended long ago. Almost certainly they are due to gases released from below the lunar surface and disturbing dust in the uppermost layer (the "regolith"). In 1960 I was Director of the BAA Lunar Section; I organized a network of observers (which still functions) and we obtained some interesting results. TLP seem to be common-est near the time of perigee, when the crust of the Moon is under maximum strain, though they may occur at any time. They are found principally in areas rich in rills, and around the peripheries of the circular maria. The most celebrated TLP-prone area is Aristarchus, the brightest crater on the Moon, which unwary observers have often mistaken for an erupting volcano (even the great Sir William Herschel fell into this trap!). Another TLP area is the floor of Gassendi, where there are many rills.

The whole question was finally settled when, in 1999, the eminent French astronomer Audouin Dollfus, using the 33-inch refractor at Meudon (Paris) detect-ed and photographed glows inside the crater Langrenus. He wrote: "They are apparently due to dust-grain levitations above the surface, under the effect of gas escaping from the soil. The Moon appears as a celestial body which is not totally dead." This, of course, is what the amateurs had been saying for years.

Following Kozyrev's observation of a red glow in Alphonsus, many of us thought that TLP would tend to be red. In fact this is not a general rule – far from it – but to search for red events we developed what became known as Moon-Blink devices. A Moon-Blink used with an adequate telescope is simply a rotating wheel, with red and blue colour filters, placed just on the object-glass side of the eyepiece (or the mirror side, with a reflector). A red patch on the Moon will show up as a dark feature when observed through a blue filter, but will be masked with a red filter.

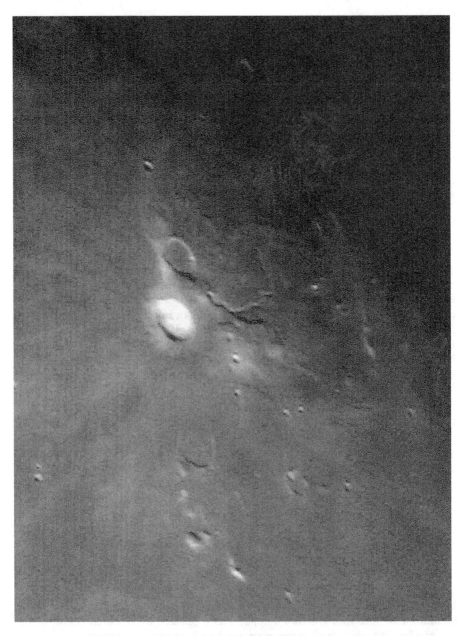

Fig. 6.8. The brilliant lunar crater Aristarchus; next to it is the darker floored Herodotus. Note the winding valley extending from Herodotus. I used my 15-inch reflector at Selsey.

Fig. 6.9. Great lunar craters: Petavius, Vendelinus and Langrenus.

Rotating the wheel, one observes first through the red, then through the blue filter in quick succession, so that any red patch will show up as a "blinking spot". The method is sensitive, and phenomena can be detected which could otherwise be missed, though naturally it is confined to red events.

It is not hard to make a Moon-Blink device, and it can be used with a modest telescope, though I would not be happy with anything smaller than an 8-inch. Yet here, above all, it is vital to avoid jumping to conclusions. It is only too easy to "see" what one half-effects to see, and a bad observation is worse than useless; it is actively misleading. I would never put much faith in a TLP observation unless supported photographically or else confirmed by another observer at a different site using a different telescope. Activity is very mild, and so are the moonquakes recorded by the instruments left of the surface by the Apollo astronauts; it is very significant that the tremors seem to be concentrated in areas where TLP have been recorded. Moonquakes pose no threat to a future Lunar Base. But what about cosmical bombardment?

The Earth's atmosphere burns up incoming meteors, which are cometary débris and are of sand-grain size. The lunar atmosphere is absolutely negligible, so that from the surface shooting-stars will not be seen. In the pre-Apollo period, many observers in the United States reported streaks across the Moon's surface which they attributed to meteoric phenomena, but it seems certain that these observations were erroneous. There were also reports in 1999 and 2000, when the Earth and Moon ploughed through the Leonid meteor stream; it was claimed that some observers using fair-sized telescopes had seen flashes on the Moon due to impacting meteors. On 18 November 1999 Brian Cudnik, using his 14-inch reflector, reported a flash as bright as a fourth-magnitude star, and a flash was also recorded on video by David Dunham. Yet a body as tiny as a meteor could not produce a flash visible from Earth. The impactor would have to be much larger – in fact, of meteorite size – and meteorites are not associated either with comets or with shooting-star meteors.

No large craters have been formed in recent times. In 1178 a monk, Gervase of Canterbury, reported seeing the Moon "split in two.... the body of the Moon writhed and throbbed like a wounded snake", and it was later suggested that the effect was due to the formation of a crater on the Moon's far side (the crater now named Giordano Bruno). But this is frankly absurd; at the time of Gervase's observation the Moon was only two degrees above his horizon, and we must dismiss the report as a "Canterbury Tale". However, the formation of a new, small impact crater is a distinct possibility.

On 15 April 1948 F.H. Thornton, a very experienced observer using a fine 9-inch reflector under good conditions, saw "a minute but brilliant flash of light" on the floor of the 60-mile crater Plato. He said that it resembled "the flash of an AA shell exploding at a distance of about ten miles". Was this due to an impact – and if so, could we locate the crater? We looked, but found nothing. Any new crater would have to be too small to be within the range of our telescopes.

On 15 November 1993 Leon Stuart, an American amateur, saw and photographed "a very bright spot on the illuminated part of the Moon". It was close to a small crater, and there were suggestions that it had been formed by the meteorite producing the flash. However, the craterlet was then identified upon a photograph taken many years ago, so that there was no possible connection. Up to now we have not traced any new impact crater.

Changeless though the Moon may be, it is surely the most magnificent of all celestial objects. There is nothing else like it, and using a telescope to explore the mountains, the valleys and the craters provides a source of endless enjoyment.

Fig. 6.10. The Mare Humorum. The large crater at the lower northern end of the Mare is Gassendi; on the edge of the Mare, to the south, note by 'bay', Doppelmayer, with its levelled north wall and its central peak. (Photographed by Commander H.R. Hatfield on 3 March 1966, with his 12-inch reflector.)

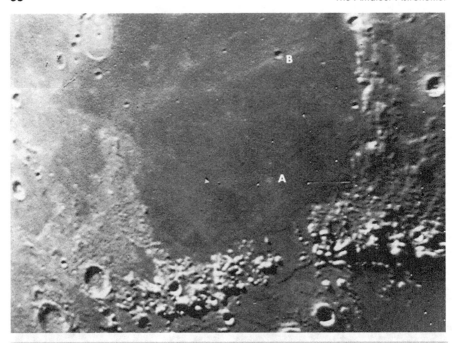

Fig. 6.11. Mare Serenitatis, as I photographed it with my 15-inch reflector. Linné is the white patch below the Mare centre (A); the upper craterlet on the Mare is Bessel (B).

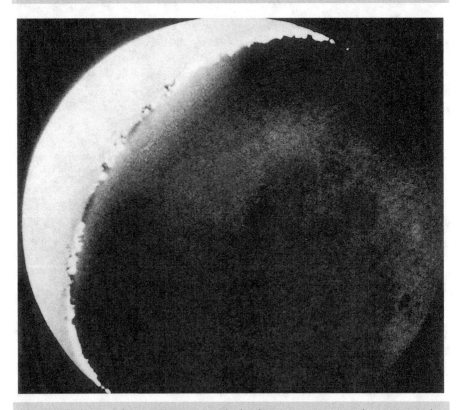

Fig. 6.12. The Earthshine, 1967 May 13. The bright crescent is necessarily over-exposed. (Photographed by Commander H.R. Hatfield.)

Chapter 7

Occultations and Eclipses

So far as the Solar System is concerned, there are long periods in which the observer has to content himself with purely routine work. This may well be followed by a number of interesting phenomena that occur in quick succession. There is perhaps a violent outbreak of sunspots, or a favourable opposition of Mars; a bright comet may make a dramatic and unexpected appearance. Also to be considered are occultations and eclipses, which have been described as the celestial equivalent of hide-and-seek.

The Moon is much the closest body in the sky, and so moves across the starry background at a relatively high speed. Sometimes it must, of course, pass in front of a star and hide it. These "occultations" are common enough, but are not so numerous as might be thought. People tend to over-estimate the size of the Moon in the heavens, and artists will usually draw it as large as a dinner- plate, whereas the apparent diameter is actually the size of a 1-inch disk held 9 feet away from the eye.

Consequently, the Moon does not pass in front of several bright stars per hour. For example, in 2003 there were only four stars above the fourth magnitude which were occulted: Epsilon Geminorum (17 January), Kappa Geminorum (18 January), Eta Leonis (19 April) and Alpha Librae (21 June). The Moon, like the bright planets, keeps to the Zodiac, so that only stars close to the ecliptic can be occulted; the brightest of these are Antares, Aldebaran, Spica and Regulus. Occultations of planets can also occur at times.

A planet shows a definite disk, so that it takes some seconds for the oncoming limb of the Moon to pass right over it. A star, however, appears as a tiny point of light, and the disappearance is virtually instantaneous. The star shines steadily until the moment of occultation, and then seems to snap out like a candle-flame in the wind. One moment it is there, the next it has gone. This is one proof that the Moon has little or no atmosphere, since a blanket of air around the limb would make the star flicker and fade for some seconds before vanishing.

Seen in a telescope, an occultation is a fascinating sight. The star seems to creep up to the Moon's limb, though actually the Moon's own motion is responsible, and the inexperienced observer is bound to feel that the star hangs close to the limb for a long time. Then the brilliant point of light will "softly and suddenly vanish away", like the hunter of the Snark, and a watcher who blinks his eyes at the wrong moment may easily miss the disappearance. The emersion, at the far limb, is equally abrupt.

Occultations used to be regarded as very important in the days when the Moon's position in the sky could not be predicted with absolute precision. The star positions were much more certain, and so the exact moment of the occultation of a star gives the position of the Moon's limb at that instant. It is true that things are different now, but occultations are still worth timing, and this is work that an amateur can do. Great care is needed, and a really good stop-watch is essential.

When an occultation report is drawn up, the following data should be added: name or number of star, time of occultation, latitude and longitude of observing station, height above mean sea level of observing station, atmospheric conditions, and any peculiar appearance seen. Occasionally, a star is hidden by a lunar peak on the limb and then reappears briefly in the adjacent valley before vanishing once more, so that it seems to wink. Timings are valuable here, because we have to take into account the jagged, irregular nature of the Moon's limb. Patience and practice

Fig. 7.1. Photo of the occultation of Saturn, 16 April 2002.
(T. Wright, with my 15-inch reflector.)

are vital in all occultation work; a small telescope can be used, provided that it is set up on a really firm mounting.

Planets can be occulted, and here of course both immersion and emersion are gradual, because a planet shows a definite disk. On 16 April 2002 the Moon occulted Saturn; I was observing it with my 15-inch reflector, together with Tim Wright, and it was most interesting. The photograph shown here (actually taken by Wright) shows how small Saturn appeared relative to the Moon. Its apparent diameter was much less than that of a large lunar crater (Fig. 7.1).

Planets, too, can occult stars. For example, on 7 July 1959 Venus occulted the bright star Regulus, and the star dimmed markedly just before and just after the actual occultation, because its light was then coming to us after having passed through Venus' extensive atmosphere. (I saw this very well, using a 12-inch reflector.) Even far-away Pluto can occult stars, and these phenomena have been used to study Pluto's tenuous atmosphere, but powerful equipment is needed.

Occultations are interesting, but eclipses are genuinely spectacular, and are bound to excite the interest even of the non-scientist. A solar eclipse is merely an occultation of the Sun by the Moon, but a lunar eclipse is very different, since the Moon is not hidden by any solid body, but passes into the cone of the shadow cast by the Earth.

The principle is shown in Fig. 7.2. The Moon has no light of its own, so that when it enters the Earth's shadow it turns a dim, sometimes coppery colour. The main cone, shaded in the diagram, is known as the umbra*, while to either side of it is the penumbra, caused by the fact that the Sun is a disk and not a sharp point of light. The diagram is not, of course, to scale, but it does serve to show what happens.

If the Moon passes right into the umbra, the eclipse is total. Every scrap of direct sunlight is cut off, but some of the Sun's rays will still reach the Moon, as they are bent or "refracted" on to it by the Earth's mantle of atmosphere, as is shown by the dashed line in the diagram. The result is that instead of vanishing completely, the Moon can usually be found without difficulty even with the naked eye. However, all eclipses are not equally dark. In 1761 the Moon disappeared so completely that it could not be seen at all, whereas in 1848 the totally eclipsed disk still shone so brightly that many people refused to believe that an eclipse was in progress. These

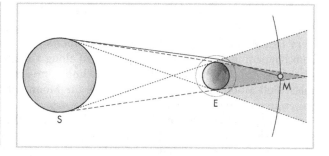

Fig. 7.2. Theory of a lunar eclipse. S, Sun; E, Earth; m, the position of the Moon at mid-totality. The diagram is not to scale.

---

*As used here, the terms "umbra" and "penumbra" have of course no connection with the umbra and penumbra of sunspots.

variations have nothing to do with the Moon itself, but are due solely to the changing conditions in our own atmosphere. It seems for instance that dust in the upper air in 1950, while vast forest fires were ranging in Canada, caused the September eclipse of that year to be rather darker than usual. Also very dark was the eclipse of June 25, 1964, when, from Sussex, I lost the Moon during totality even with my 12½-inch reflector, though conditions were not ideal. The cause on this occasion was volcanic dust which had been sent into the upper atmosphere by an earlier eruption in the East Indies. By the eclipse of the following December, much of this dust had settled, and the eclipse was lighter, though still rather dark by normal eclipse standards.

I have fond memories of one total lunar eclipse, that of 24 April 1986. Halley's Comet was fading as it moved away from the Sun and the Earth; it was dropping below naked-eye visibility, but when the Moon entered eclipse the comet was well seen. I happened to be on the island of Bali at the time, and the effect was truly magical. It will not be until the year 2061 that Halley's Comet will again be visible with the naked eye.

The French astronomer André Danjon introduced an 'eclipse scale', which is distinctly useful. There are four grades:

1. Dark eclipse with greys and browns, with surface details barely identifiable.
2. Deep red or rusty, with the outer edge of the umbra relatively bright.
3. Brick-red, with a bright yellow rim to the shadow.
4. Very bright; orange or coppery red, with a bright bluish shadow rim.

Danjon tried to link eclipse brightness with the state of the sunspot cycle, but it must be admitted that the results were rather inconclusive.

If the Moon does not enter wholly into the umbra, the eclipse is partial, while at other times it is merely penumbral. Penumbral eclipses will not be noticed except by the attentive watcher, since the dimming is too slight to be conspicuous.

Two things are clear from the diagram. First, a lunar eclipse must be visible from one complete hemisphere of the Earth, provided that clouds do not conceal it, and if it is total anywhere it must be total everywhere. Secondly, an eclipse can happen only at Full Moon. If the Moon passes through the centre of the umbra, it may remain totally immersed for over an hour, while the partial phases can extend over four hours.

Lunar eclipses are so obvious that they must have been known from early times. Were the Moon's orbit not tilted across the ecliptic, a total eclipse would happen at every Full Moon, but the inclination of the Moon's path is enough to prevent this from happening. Imagine two hoops hinged along a diameter, and crossing each other (Fig. 7.3). The points at which the two hoops cross are called the "nodes", and unless Full Moon occurs very near a node the Moon will miss the shadow altogether, so that no eclipse will occur.

It so happens that the Sun, Moon and node return to the same relative positions after a period of 18 years 10¼ days, and so any particular eclipse will be followed by another eclipse 18 years 10¼ days later. This is the so-called Saros Period, and was used by the Greek astronomers to make eclipse predictions. The Saros is not exact, but the method is accurate enough to be workable. A list of future eclipses is given in Appendix 9.

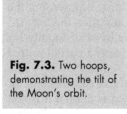

Fig. 7.3. Two hoops, demonstrating the tilt of the Moon's orbit.

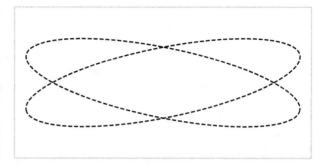

Eclipses are always fascinating to watch, but can there be any detectable effects on the Moon itself? Since there is virtually no atmosphere to protect the surface, and since the upper layer ("the regolith") is bad at retaining heat, the temperature drops as soon as the eclipse begins. The fall may amount to 100°F in the course of an hour and there have been claims that certain formations show changes. Linné on the Mare Serenitatis, consists of a craterlet surrounded by a white nimbus, and it was suggested that this nimbus brightened up during an eclipse. I am frankly sceptical, and my own observations have been negative, but I suppose that it is worth monitoring formations such as Linné, Plato and Aristarchus.

Turning now to eclipses of the Sun, we find that principles involved are very different. We are back to the "occultation" idea, since a solar eclipse is caused simply by the Moon passing between the Sun and Earth.

The Moon is far smaller than the Sun, but it is also so much nearer that in our skies it looks almost exactly the same size. When the three bodies move into a straight line, with the Moon in the middle, the shadow cast by the Moon just touches the Earth's surface, and for a few minutes the bright solar disk is blotted out by the dark, and therefore invisible, body of the Moon (Fig. 7.4). The width of the completely shadowed area of the Earth is 167 miles at best, and so a total solar eclipse will not be seen over a complete hemisphere; for instance, the eclipse of 11 August 1999 was total as seen from Plymouth, but only partial as seen from London.

To either side of the track of totality, the eclipse will be partial, while some eclipses are not total anywhere on the Earth. There is also a third kind of eclipse, the annular (Latin *annulus*, a ring). As we know, the Moon moves in an elliptical path, so that its distance from the Earth varies. When at its most remote, it appears smaller than the Sun in the sky, and so cannot cover the whole of the solar disk.

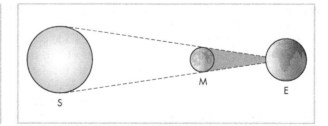

Fig. 7.4. Theory of a solar eclipse. S, Sun; M, Moon; E, Earth. The diagram is not to scale.

When the three bodies line up under these conditions, a bright ring of the Sun is left showing round the dark side of the Moon.

Obviously, a solar eclipse can happen only at New Moon, and then only if New Moon occurs near a node. The Saros period is valid, as for lunar eclipses, but the rough and ready method of forecasting is less accurate. For instance, the "return" of the 1927 eclipse took place in 1945, but in this year the band of totality lay further to the north, and missed England altogether, so that only a partial eclipse was seen in our country.

Partial and annular eclipses are spectacular, but do not give much scope for useful work. Remember, too, that even when most of the Sun is hidden it is still unsafe to use direct vision either with binoculars or with a telescope. The slightest sliver of sunlight remaining is enough to damage the eye in a matter of seconds.

A total of eclipse is among the grandest of Nature's displays. As the Moon sweeps on, the light fades, until at the instant the last of the disk is blotted out the atmosphere of the Sun leaps into view. There are the magnificent prominences; there is the glorious chromosphere, and there too is the "pearly crown" or corona, a superb glow surrounding the eclipsed Sun, sometimes fairly regular in outline and sometimes sending out streamers across the heavens. It is a pity that the spectacle is so brief. No total eclipse can last for more than about 8 minutes, and most are much shorter, so that the astronomers are ready to travel to remote parts of the world in order to make the most of their limited opportunities. This enthusiasm is not merely for the beauty of the sight; there are many investigations that cannot be made except during the period of totality and the few seconds before and after. In fact, serious workers are so busy that they have no time to stop and admire what is going on.

The prominences are visible to the naked eye only during totality, but with special instruments they can be seen at any time. They are made up of incandescent gas, and are of tremendous size; the length of an average prominence has been given as 125,000 miles. Many are associated with sun-spots, and prominences too are affected by the 11-year solar cycle.

"Quiescent" prominences are relatively calm, as their name suggests, and they may last for several months before either breaking up gradually or being violently disrupted. Active prominences may be likened to tall tree-like structures, from the tops of which glowing streamers flow out horizontally and then curve downwards towards the bright surface of the Sun. Some of these active prominences are truly eruptive, and it has been known for the blown-off material to move at over 400 miles per second.

Prominences can be seen at any time by using filters which block all the light except that emitted by hydrogen and these are now available to all amateurs who are interested in the Sun. Films taken of eruptive prominences are truly dramatic.

The Sun's outer atmosphere – the corona – is much more difficult to observe from Earth except during a total eclipse. It is made up of very tenuous gas, millions of times less dense than the Earth's air at sea level, and it stretches out for a vast distance, though because of its tenuity and its indefinite boundary it is not possible to give an exact figure for its "depth".

Seen from any one place on Earth, total eclipses are rare; from England the last two were in 1927 and 1999, and for the next we must wait until 2090, but other parts of the world are more favoured; a list of forthcoming eclipses is given in the

Appendix. Partial eclipses are much more common, but are really no more than of casual interest; annular eclipses can be quite striking, and if the Sun is almost covered it is sometimes possible to glimpse a prominence or two, but even an annular eclipse is a very poor substitute for totality.

Remember, it is safe to look direct, with a telescope or an SLR camera, only when totality is complete, and the corona is on view. At any other time the Sun is just as dangerous as it is when uneclipsed.

I have been lucky enough to see six totalities. My first was on 30 June 1954, and I think it worth describing here, simply because it was entirely new to me – and first impressions are often the best.

The track of totality brushed the northernmost of the Shetland Islands, off the coast of Scotland, and so a large partial eclipse was seen in England, and caused general interest even among people not usually astronomically-minded. The main track crossed Scandinavia, where many astronomers gathered. The combined Royal Astronomical Society and British Astronomical Association party, of which I was a member, made its headquarters at the little Swedish town of Lysekil, along the coast from Göteborg, since weather conditions in West Sweden were expected to be rather better than in Norway (as did indeed prove to be the case). Our arrival in Lysekil coincided with the Midsummer Festival. It also coincided with a burst of torrential rain.

On June 30, most observers collected their equipment and drove to Strömstad, in the exact centre of the track, almost on the Norwegian frontier. The site selected was a hill overlooking Strömstad itself, and by noon it was littered with equipment of all kinds; telescopes, spectroscopes thermometers, cameras, and even a large roll of white paper that I had spread out in the hope of recording shadow bands. These shadow bands are curious wavy lines which appear just before totality. They are due to atmospheric effects, but have seldom been properly photographed; the opportunity seemed too good to be missed.

The early stages of the eclipse were well seen. Five minutes before totality, everything became strangely still, and over the hills we could see the approaching area of gloom. Then, suddenly, totality was upon us. The corona flashed into view round the dark body of the Moon, a glorious aureole of light that made one realize the inadequacy of a mere photograph. The sky was fairly clear; and although a thin layer of upper cloud persisted, only those with the experience of former eclipses could appreciate that we were not seeing the phenomenon in its full splendour.

It was not really dark. Considerable light remained, and of the stars and planets only Venus shone forth. Yet the eclipsed Sun was a superb sight indeed, with brilliant inner corona and conspicuous prominences. The two and a half minutes of totality seemed to race by. Then a magnificent red-gold flash heralded the reappearance of the chromosphere; there was the momentary effect of a "diamond ring", and then totality was over, with the corona and prominences lost in the glare and the world waking once more to its everyday life. In a few minutes, it was almost as though the eclipse had never been.

All the totalities I have seen have had their own special features. The shape of the corona varies according to the state of the sunspot cycle; sometimes there are magnificent prominences, at other times almost none; sometimes the planets and bright stars shine out, while at other times they do not. All in all I think that the loveliest of all 'my' eclipses was that of 24 October 1995, which I saw from the deck of the cruise ship, the *Marco Polo*, in the China Seas. On this occasion I was not making a live television programme and Chris Doherty, who was with me, was taking the photographs, so that I was at liberty to do nothing except watch. It was breathtaking. The sea was calm – our Norwegian captain has manoeuvred us to exactly the right place and I saw the Moon's shadow racing across the water; then the darkness fell, and I had the feeling of having been transported to another planet. The corona was superb, and there was a huge red prominence. Yet I noted that one dear lady, sitting in a deckchair some way from me, continued to read throughout totality. It must have been an interesting book.

I was less lucky on 11 August 1999, from Falmouth in Cornwall. I had a BBC team with me, plus two very close astronomical friends, Peter Cattermole and Iain Nicolson. The preceding morning was brilliantly clear, but on eclipse day the sky was completely overcast, and all we could do was to crouch under a large umbrella, muttering the equivalents of "Tut, tut! Dear me! How annoying!" The eerie gloom was most impressive, but the Sun refused to emerge (Fig.7.6).

May 30, 2003, provided an annular eclipse visible from North Scotland. Actually the best place from which to view it was Iceland, but the BBC wanted me to show the eclipse on television and could not afford to send a team beyond Britain, so I went to the Elgin area, with Dr Steve Wainwright, Iain Nicolson and Dr Brian May (yes, the great guitarist; many people forget that he is a highly qualified astronomer). Despite cloud trouble, we saw the "ring of fire", though of course

Fig. 7.5. A total eclipse of the Sun, showing the corona. (Photograph by Chris Doherty.)

Fig. 7.6. Total solar eclipse, India, October 1995. (Photographed by Michael Maunder.)

there was no sign of the corona or prominences, and it is true that an annular eclipse seems very tame when compared with totality (Fig. 7.7).

If you ever have the chance to see a total eclipse, do not hesitate. In one way we are very fortunate. It is sheer chance that the Sun and the Moon appear so nearly equal in the sky; were the Moon a little smaller, or a little further away, the solar corona might still remain unknown.

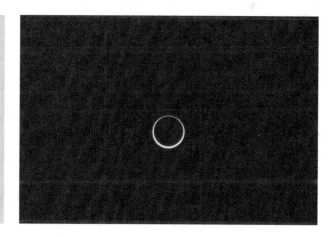

Fig. 7.7. Annular eclipse of the Sun, 1994. (Photographed by the author.)

Chapter 8

Aurorae and the Zodiacal Light

It is impossible to separate once science from another. Even astronomy is no longer "on its own", as it used to be. It is bound up closely with chemistry and physics, and it is also linked with weather study, or meteorology, by the pheno-menon known as the Aurora Polaris, or Polar Light.

Aurorae have been known from very early times, and are so common in high latitudes that a night in North Norway or Antarctica would seem drab without them. In England they are less frequent, though displays are seen on an average at least ten times a year, while in the tropics they are rare. They are not unknown; there is a famous story of how the cohorts of the Roman emperor Tiberius once rushed northwards to the help of the people of Ostia because of a red glow in the sky that they took for a tremendous fire, but which proved to be merely an aurora. However, there can be no doubt that observers in the far north and south have the best views.

Aurorae occur in the upper atmosphere, at heights ranging from over 600 miles down to as low as 60. Sometimes the lights take the form of regular patterns, while at others they shift and change rapidly, often showing brilliant colours and provid-ing a spectacle that is second only to the glory of a total solar eclipse. One of the greatest displays of modern times took place on January 25, 1938, when all Britain witnessed the spectacle. From Cornwall "the whole of the western sky was lit with a vivid red glow like a huge neon sign; gradually shafts of white light were inter-mingled with the redness, changing quickly to an uncanny grey light and then to a brilliant silver, while green patches appeared here and there". From Sussex the dominating colour was red, though during the course of my own observations I recorded many other hues as well. The aurora was brilliant and widespread enough to cause interest and even alarm over the whole country, and it was seen from places as far south as Vienna (Fig. 8.1).

So far as England is concerned, there have been few comparable displays since then, but there were brilliant aurorae on 13 March 1989 and 8-9 November 1991; there were vivid hues, and from much of England the aurora cast obvious shadows. There was another good display on 6 April 2000, when again the predominant colour was red. But of course England is not in the "auroral zone". On average, aurorae are seen on 240 nights per year in North Alaska, Iceland and North Norway; 25 nights per year in central Scotland and along the Canada-USA border,

Fig. 8.1. Aurora Borealis, seen by the Yukon River-Circle, Alaska , in August 2000. (Photograph by Dirk Obudzinski.)

and only one night per year in Central France, while Greece and Italy may have no more than one or two good displays per century. The Southern Lights are of course on display from Antarctic, to be enjoyed mainly by the local penguins, but are sometimes seen from the South Island of New Zealand and even the southernmost tip of Africa.

Since meteorology is the science of the atmosphere, and aurorae are atmospheric phenomena, it might seem logical to assume that they are not the concern of astronomers. Yet the cause of aurorae is to be found not on the Earth, but in the Sun. Active regions on the Sun send out electrically charged particles, which stream across space to the Earth and enter the upper air, making it glow. Because the particles are electrified, they spiral down toward the magnetic poles, though the process is not quite so straightforward as might be thought. Auroral activity is more or less permanent at high latitudes (both north and south) along the so-called auroral ovals, which are "rings" symmetrically displaced around the magnetic poles. Thus from Tromsø in Norway, latitude 69 degrees north, aurorae are commoner than they are as seen from the North Pole itself. When the streams of particles are particularly intense, the ovals broaden and expand, bringing auroral displays further north and south of the main regions.

Aurorae are high-altitude phenomena; generally they lie about 60 miles from ground level, and extend up to around 180 miles, though in extreme cases they may reach 300 miles. They always lie far above normal clouds, so that with an overcast sky there is no chance of seeing an auroral display of any kind. Active regions on the Sun are often associated with spots, so that aurorae are most frequent at or near spot maximum. Moreover, a major flare occurring near the centre of the solar disk is often followed a day later by a bright aurora. Since the particles must therefore cover the 93-million-mile gap in about 24 hours, this delay indicates a speed of about 1,000 miles per second.

One interesting and puzzling problem is that of "auroral noise". Hissing sounds, rustling and whistling have been reported on many occasions, and are extremely difficult to explain. Odours have also been reported, but smelly aurorae seem to be even more baffling than noisy aurorae.

Scientifically, aurorae are important not only because of their link with the Sun, but because they provide information about the upper air. It is therefore useful to observe them whenever possible, and to make estimates of their positions against the starry background, so that their heights may be worked out. The main work here has been done by Norwegian scientists, led by Professor Carl Størmer of Oslo; but amateurs can play a major role; a full-scale survey has been organized by the British Astronomical Associations's Aurora Section, which has members all over the world. Observers taking part are asked to fill in forms telling of the presence or absence of aurora, coupled with notes of any displays that may be seen. Negative reports are not without value, and may in fact be of great help.

The lights are so varied that to describe all the forms would need more pages. One never knows quite what an aurora is going to do next, but a great display often begins as a glow on the horizon, rising slowly to become an arc. After a while, the bottom of the arc brightens, sending forth streamers, after which the arc itself loses its regular shape and develops folds like those of a radiant curtain. If the streamers extend beyond the zenith, or overhead point of the sky, they converge in a patch to form a corona (not, of course, to be confused with the corona which surrounds the Sun). Finally the display sends waves of light flaming up from the horizon towards the zenith, after which the light dies gradually away. The whole phenomenon may extend over hours.

For observing aurorae, by far the best instrument is the naked eye, coupled with a red torch and a reliable watch. Binoculars are of little help, and telescopes absolutely useless. Points to note are the bearing of the centre of the display, reckoned in degrees (0 to 360) from north round by east; the type and prominence of the aurora; the various forms seen, such as arcs, curtains, draperies and flaming surges; colours, and duration. Times should be taken at least to the nearest minute. There is obvious scope for the photographer, and spectroscopic work is of great interest, but simple naked-eye observation is not to be despised.

Though aurorae are so spectacular, they are not the only lights seen in the heavens. The sky itself seems to shine with a feeble radiance known as the airglow, and sometimes a cone of light can be seen after dusk or before dawn, extending upwards from the hidden Sun and tapering towards the zenith. Since it extends along the Zodiac, this cone is known as the Zodiacal Light. It can be quite prominent when seen from countries where the air is clear and dust-free, but from Britain it is always hard to see. The Zodiacal Band, a faint, parallel-sided extension of the cone, may extend right across the sky to the far horizon, though it is so dim that it is seldom to be observed at all except from the tropics.

Unlike the aurora, the Zodiacal Light originates well beyond the top of the air. It is due to light reflected from a layer of thinly-spread matter extending from the Sun out beyond the orbit of the Earth, rather like a tremendous plate. The layer cannot be broad, as is shown by the fact that the Light is never seen except close to the ecliptic. The best times for observations are late evenings in March and early mornings in September, because at these times the ecliptic is most nearly perpendicular to the horizon, and the Light is thus higher in the sky (Fig. 8.2).

Since the Zodiacal Light is faint, so its intensity is not easy to estimate. The best way to measure it is to compare it with a definite area of the Milky Way, and the width of the base, in degrees, should also be noted. Though the Light is predominately white, a pinkish or at least warmish glow has been reported in the lower parts, and should be looked for.

Last and most elusive of these glows is the Gegenschein, which is a faint hazy patch of light always seen exactly opposite the Sun in the sky. It appears at its most conspicuous in September, when it looks like a round luminous patch about forty times the apparent width of the Moon, but it is extremely hard to see, and even a distant lamp is enough to hide it. From England I have looked for it frequently, but have seen it only once in 1942, when the whole country was blacked out because of German air-raids. It is excessively difficult to photograph, though with the latest equipment some good pictures have been obtained. The German name is generally used; the English equivalent is the "Counterglow".

For all these observations, one thing should be borne in mind: Never begin work before you have made your eyes thoroughly accustomed to the dark. To come outdoors from a brilliantly-lit room and expect to see an auroral glow or the Zodiacal Light straight away is fruitless, and it is usually necessary to walk about

Fig. 8.2. Aurora in Boötes, March 3 1989 (Photograph by Paul Doherty.)

for at least a quarter of an hour before starting your programme, though the exact period is bound to vary with different people. For recording observations, a torch with a red bulb is the ideal, since an ordinary white light will dazzle you sufficiently to ruin the sensitivity of your eyes for some minutes afterwards.

Here again, then, the amateur has a part to play. There is no need to wait years for a great aurora; studying the fainter lights and glows is a fascinating hobby, and it is a pity that city dwellers never have a chance to see the ghostly beauty of the Zodiacal Light.

Chapter 9

The Nearer Planets

There was a time, not so long ago, when most amateur astronomers concentrated almost exclusively upon the members of the Solar System. This is not true today, but of course the Moon and planets remain favourite targets – and despite the space probes, there is still a great deal about them that we do not know. Even a small telescope will show details on some of them, so let us consider the planets one by one.

The four members of the inner group – Mercury, Venus, the Earth and Mars – are solid, rocky bodies. They are comparable in size, and all have atmospheres of a kind (though that of Mercury is extremely tenuous). These are the only common factors. Otherwise, they are as different as they can be.

Mercury, whirling round the Sun at an average distance of only 36 million miles, is never easy to observe. It always lies somewhat near the Sun's line of sight, and we can well understand why it was named after the elusive, fleet-footed Messenger of the Gods. Moreover it is not much larger than the Moon, and is more than 200 times as distant, so that ordinary telescopes will show little except a tiny disk with its characteristic phase.

Mercury takes 88 days to complete one orbit found the Sun. It used to be thought that the axial rotation period was exactly the same, so that Mercury would always keep the same face turned sunward, just as the Moon does with respect to the

Fig. 9.1. Comparative sizes of the Earth, Mercury and the Moon.

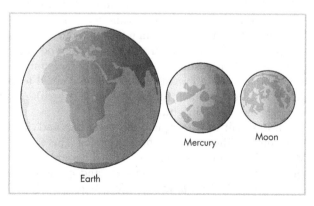

Earth

Mercury

Moon

Earth. Over part of the surface there would be perpetual daylight, with a surface temperature exceeding 700 degrees Fahrenheit; over another part there would be everlasting night, so that the surface would be colder even than remote Pluto. Between these two extremes there would be a "Twilight Zone" over which the Sun would rise and set. Mercury's orbit is not circular, and so its velocity varies between $36\frac{1}{2}$ miles per second at perihelion and only 24 miles per second at aphelion. This would result in effects analogous to the Moon's librations, so producing the Twilight Zone.

Then, however, it was found that the 'night' hemisphere is not nearly so cold as it would be if it received no sunlight at all. Radar measurements confirmed that the true rotation period is 58.6 days, two-thirds of a Mercurian year. Science fiction writers were disappointed to learn that there is no area of permanent day, no region of everlasting night, and no Twilight Zone.

Our first real knowledge of the surface features was due to the U.S. probe Mariner 10. It was launched in November 1973; in the following February it flew past Venus, using the gravitational pull of that world to direct it in towards a rendezvous with Mercury. The first active pass of Mercury was made in March 1974, and there were two more before the probe finally "went silent". Many hundreds of close-range photographs were obtained, and it became clear that superficially, at least, Mercury is very like the Moon. There are craters, mountains and valleys, and even one huge, mountain-ringed structure – now known as the Caloris Basin – which looks decidedly "lunar".

Less than half of the total surface was surveyed by Mariner 10, but there is no reason to suppose that the remaining areas are fundamentally different. Other interesting facts also emerged. Mercury has a definite magnetic field; it is much weaker than the Earth's, but it exists, and presumably indicates the presence of a large iron-rich core. There is a trace of atmosphere, but the density is so low that it corresponds to what we would normally call a vacuum, and it will be of no use whatsoever to possible astronauts of the future. It is painfully clear that so far as Mercury is concerned, any life of the kind we know is quite out of the question.

I have glimpsed a few patches on Mercury, with a 6-inch refractor, but little can be seen with amateur-owned telescopes. This does not mean that there is no point in looking for Mercury. It is always satisfying to see the strange little world glittering shyly in the late evening or early morning, and on an average it can be seen with the naked eye at least a dozen times each year.

When Mercury is glimpsed without a telescope, it is bound to be near the horizon, so that it will be shining through a deep layer of the Earth's atmosphere, and the image will be unsteady. The best method is to find the planet as early as possible, while it is still fairly high up; for sweeping, it is advisable to use either binoculars or else a low magnification on a small telescope (I have found that a power of 25 on a 3-inch refractor does very well). A drawing can then be made while the sky is still bright.

Telescopes fitted with equatorial mountings and clock drives allow a faint object to be found without any tiresome sweeping. This saves a great deal of time, though Mercury is never easy to locate except with a large instrument. Yearly star almanacs tell where and when it is to be seen, and more detailed tables are given in *The Handbook of the British Astronomical Association.*

It is also possible to sweep for Mercury in broad daylight, but it is never wise to range about with a telescope until the Sun has set. Moreover, Mercury and the Sun will not be far apart, and there is always the chance that the Sun will enter the field of view during the sweeping, with disastrous results.

For examining the phase and for drawing any visible surface markings, the magnification used should be as high as possible, but the slightest unsteadiness or blurring will be fatal, so that one has to strike a happy mean. All things considered, Mercury is more difficult to study than any other planet, and it is hopeless to expect to see anything spectacular. People who live in or near cities will be lucky to find it at all.

Mercury can sometimes pass in transit across the face of the Sun. The last transit was that of 7 May 2003, and using the projection method I was able to follow it with my 5-inch refractor. Though I should have been prepared, I was surprised to see how small Mercury looked; it was indeed tiny, and much blacker than any sunspots. The next transits will be those of 8 November 2006 and 9 May 2016.

Venus, the second of the inferior planets, is as different from Mercury as it could possibly be. It is almost the same size as the Earth, and is the nearest body in the sky apart from the Moon. It is also more brilliant than any other object than the Sun and the Moon. It is at its very best during the crescent stage, since by the time of "dichotomy" (half phase) it has already drawn away from us, so that its apparent diameter is much less (Fig. 9.2). Altogether, Venus is a most infuriating object.

Moreover, the disk appears virtually blank even with powerful telescopes. Vague, dusky shadings may be seen often enough, but they are not permanent, and are so diffuse that they are hard to define. In fact, we are not looking at the true surface of Venus at all; we are seeing only the upper layers of an obscuring atmosphere.

Before the end of 1962 we knew practically nothing about the surface of Venus, and there were constant references to "the Planet of Mystery". It was known that the atmosphere was rich in carbon dioxide, and since this gas acts in the manner of a greenhouse, the surface had to be hot – but how hot? Even a problem as fundamental as the length of the axial rotation period remained unsolved; various estimates were given, ranging from $22\frac{1}{2}$ hours up to 225 days. In the latter case the rotation would be synchronous, so that the same hemisphere would be turned toward the Sun all the time. Opinions as to the nature of the surface fluctuated

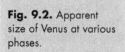

Fig. 9.2. Apparent size of Venus at various phases.

wildly between a swampy, tropical hothouse, a planet completely covered by water, and an arid dust-desert without a scrap of moisture anywhere.

Only space research methods could give us the truth. The first successful probe was America's Mariner 2, in 1962, which by-passed the planet at 21,000 miles and proved that the surface really is incredibly hot. (In case you are wondering what happened to Mariner 1, I have to tell you that it fell into the sea as soon as it was launched; apparently someone forgot to feed a minus sign into a computer, which makes quite a difference.) Other missions followed, and in October 1976 two Russian vehicles, Veneras 9 and 10, made controlled landings; each sent back one picture before being put out of action by the hostile environment. Since then the U.S. orbiting Magellan has mapped the whole surface by radar, so at last we know what Venus is like. There are uplands, lowlands and craters; there are volcanoes, presumably active, and lava-flows everywhere. The atmosphere is almost pure carbon dioxide, with a ground pressure 90 times that of the Earth's air at sea-level; the surface temperature is nearly 1000 degrees Fahrenheit, while the clouds contain large amounts of sulphuric acid. Obviously, it is not a good place to visit for a peaceful weekend. There is no detectable magnetic field; the rotation period is 243 days – longer than Venus' "year" – and the planet spins from east to west, in a sense opposite to that of the Earth. The chances of finding life there are effectively nil. So what can telescope studies from Earth tell us?

There is no point in observing Venus when it is shining brilliantly down from a dark sky; the disk will be dazzling, and the image is likely to be violently unsteady. I have found that the best seeing is obtained when the planet can just be detected with the naked eye, shortly after sunset or shortly before sunrise, but observations made in broad daylight are almost as good. Venus is so bright that it can usually be found without much difficulty even when the Sun is above the horizon. In general it will not stand a high magnification, but I have often used over ×300 on my 15-inch reflector.

It is worth trying to draw the vague dark patches, and to see how they shift, but well-defined markings are very rare. Bright areas are usually seen near the poles; they are cloud formations and so are completely different from the white caps of Mars. But there is another line of research which is interesting to follow up. This involves the exact moment of "dichotomy", or half-phase.

Since the orbit of Venus is known so accurately, it should be easy to predict the time of dichotomy, but these predictions are usually wrong by several days. When Venus is an evening object, observed dichotomy is always early; with morning elongations, dichotomy is late. The first astronomer to notice this curious disagreement was Johann Schröter, and in writing a paper about it years ago I called it Schröter's Effect", a term which now seems to have been generally accepted.

There is no chance of being out of position, and the effect must be due to the planet's atmosphere. Timing the actual date of dichotomy is therefore valuable. The terminator will appear sensibly straight for several nights in succession, so that a series of observations is necessary. What generally happens, of course, is that clouds intervene at a critical stage, and cause one to miss a vital evening's observation. Filters can be a help, and it is worth using several in turn to check against observations made in ordinary light. (Mercury, incidentally, does not seem to show a Schröter effect, which is understandable in view of its lack of atmosphere. More measurements are needed, but telescopes of considerable power are required, and for this sort of work the refractor has the advantage over a reflector.)

The terminator of Venus shows occasional slight indentations and projections. Schröter believed them to be due to differences in level, and thought that he had charted a mountain 87 miles high (!), but we may now be sure that cloud effects are responsible.

Last, but by no means least, there is the Ashen Light. When Venus is crescent, the night area can often be seen shining faintly, so that the whole disk can be traced. The same appearance can be seen with the crescent Moon, but the cause is different. With the Moon, the glow is due to reflected earthlight, but the Earth is certainly unable to illuminate Venus, and Venus itself has no moon. Weird theories have been advanced to explain the Ashen Light – in 1840, Gruithuisen suggested that it might be due to general festival illuminations lit by the inhabitants of Venus to celebrate the crowning of a new ruler, but since the Light is by no means unusual it would seem that the Government of Venus would be as unstable as that of an African banana state! There have been claims that it is a mere contrast effect, but I am confident that this is not so. I made a special telescope eyepiece with a curved occulting bar, which would cover the bright crescent – and the dimly-lit "night" side could still be seen.

One interesting theory is that it is caused by brilliant aurorae in the upper atmosphere of Venus. Since Venus is closer to the Sun than we are, there is no reason to doubt that aurorae exist, and the explanation is not unreasonable. If we could show that the Light is at its brightest during periods of solar activity, when terrestrial aurorae are frequent, we might be able to clear up the problem. I have made an attempt to analyse the available observations of the Light, but unfortunately the observations themselves are too scattered to be of much use. It must however be added that the "aurora" theory has been weakened by the revelations that Venus has no appreciable magnetic field.

All the markings of Venus are so indefinite that they are hard to record, and there is the added complication that the great brilliancy of the disk tends to produce regular, streaky patterns that do not actually exist at all. The nebulous aspect should be drawn as faithfully as possible, and if the depth or sharpness of a shading is exaggerated – as is sometimes necessary – the observer must always be careful to write an explanatory note beside his sketch. A scale of 2 inches to the planet's diameter is convenient.

Venus, like Mercury, can pass in transit across the face of the Sun, but it does so much less often. Transits occur in pairs separated by eight years, after which there are no more for over a century. The last transits were in 1874, 1882 and 2004; the next will be on 6 June 2012, after which we must wait until 11 December 2117. The transit of 8 June 2004 was well seen from my observatory (see Figure 9.3, overleaf) – about a hundred observers joined me there! – and Venus was very evident even with a 'pinhole camera'. We had a splendid view of the Black Drop which had so frustrated earlier observers. Let us now turn to Mars, which is much more rewarding inasmuch as it does show permanent surface markings.

Mars can approach us to a distance of 35 million miles. It is therefore always at least 150 times as remote as the Moon, with a diameter of only 4,200 miles, and it is much smaller than the Earth. Fortunately, it is better placed than either Mercury of Venus. Since it lies beyond the Earth's orbit, it can never appear as a half or a

Fig. 9.3. Transit of Venus, 8 June 2004, as seen from Selsey, UK.

crescent, while at its most gibbous it looks like the shape of the Moon two or three days from Full.

The main trouble about observing Mars is that it comes to opposition only at intervals of nearly two years, as explained in Chapter 4. Things are at their best for only a few weeks before and after the opposition date, so that the observer has to make the best use of the limited time at his disposal. Not all oppositions are equally favourable, because the orbit of Mars is much more elliptical than ours, and the best oppositions occur when Mars is near perihelion. The opposition of 2003 was much better than that of, say, 2010 (see Figure 9.4). At the end of August 2003 Mars was slightly less than 35,000,000 miles from us – as close as it can ever be. For a few weeks Mars outshone even Jupiter, but when the planet is at its furthest from Earth it is not much brighter than the Pole Star.

Mars has a "year" of 687 Earth days (668 Mars days or 'sols'), and the axial inclination is almost the same as ours, so that the seasons are of the same type. Since the mean distance from the Sun is about 48 million miles greater than that of the Earth, we must expect it to be cool; the maximum summer temperature at the equator can be as high as 50 degrees Fahrenheit, but the nights are much colder than a polar

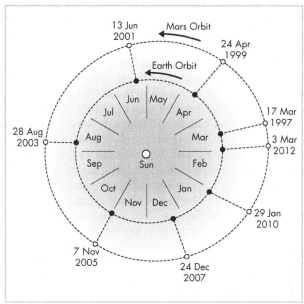

Fig. 9.4. Oppositions of Mars 1997–2012.

night on Earth. The atmospheric pressure at ground level is below 10 millibars everywhere, and the main constituent is carbon dioxide.

The newcomer to astronomy is apt to be disappointed with his first telescopic view of Mars. Generally there is little to see apart from a tiny disk, reddish in colour, crowned in the north or south by a whitish cap. Seeing much on Mars requires a good deal of practice, because the markings are much less spectacular than the belts of Jupiter, the rings of Saturn or even the phases of Venus. The polar caps, too, are very variable in extent. Moreover, global dust-storms sometime develop which hide the surface details completely.

In 1877 the Italian astronomer G.V. Schiaparelli drew a detailed map of Mars, using a fine 9-inch refractor. He charted the bright areas and the dark patches, and gave them names, most of which are still in use – but he also drew straight artificial-looking lines which he called "canali". This is Italian for "channels", but inevitably it was translated as "canals". What were they?

Schiaparelli kept an open mind, but the American astronomer Percival Lowell did not. He was convinced that the canals were artificial, built by the "Martians" to pump water from the polar snows through to the centres of population in warmer latitudes. Lowell set up an observatory at Flagstaff, in Arizona, and equipped it with a fine 24-inch refractor. From 1895 he produced drawings which showed an intricate canal network, and he also claimed that some of the canals became abruptly double, so that presumably the engineers there opened up new waterways.

Even in Lowell's lifetime there was a good deal of scepticism, but others beside Lowell drew the canals, and almost everyone believed that the dark areas were covered with vegetation. Moreover, one thing was definite. If the canal drawings were accurate, then Mars would be inhabited. A network of that kind could not develop naturally. Yet some observers, using telescopes as powerful as Lowell's, could not see the canals at all, or else drew them as disconnected patchy strips.

There was the problem of the "wave of darkening", bound up with the regular cycle of the polar caps, which wax and wane according to the seasons. It was claimed that when a cap shrank, the dark areas near its border became sharper, as though the vegetation were being revived by moisture in the atmosphere. It seemed credible enough even to those who had no faith in Lowell's Martians.

I well remember my first views of Mars through Lowell's telescope, in the late 1940s. Would I see canals, and follow the wave of darkening? I did not, and am very glad that I failed, because the canals do not exist; they were mere tricks of the eye. It is only too easy to "see" what you half expect to see. Much later, I superimposed Lowell's canal network on a modern map of Mars, and found that there was no correlation with any genuine surface features.

From 1965 came the flights of the space-probes, and all our ideas about Mars had to be revised. There are mountains, valleys and craters; there are also huge volcanoes, which may or may not be mildly active. The highest, Olympus Mons, towers to 15 miles, and has 40-mile caldera at its summit. (I remember seeing it with my modest 15-inch reflector, but only as a tiny speck, and I had no idea of its true nature.) There are also basins, such as Hellas, which can be cloud-filled and then masquerades as an extra polar cap. One striking feature is the Mariner Valley, which is 2000 miles long and up to five miles deep, with raised, ragged rims; it makes our Grand Canyon of the Colorado seem very puny. The Martian landscape is nothing if not dramatic, and most people will have seen the pictures sent back by space-craft which have made controlled landings there.

Is there life on Mars? There may be – perhaps surviving in underground lakes or seas, but we can expect nothing more advanced than tiny, single-celled organisms. There can be nothing so elaborate as a blade of grass.

In view of all that we have found out from the space-probes, is there any point in looking at Mars through a telescope, apart from the sheer enjoyment of it? The answer is: "Yes". Mars has atmosphere, and not a static world. Studies of dust-storms are useful, and the same is true of the ice-crystal clouds, which shift and tell us a great deal about the Martian wind systems. We can also follow the changes in the polar caps. (Note, incidentally, that they do not 'melt'; liquid water cannot exist on the surface of the Mars, because the atmospheric pressure is too low. A shrinking cap "sublimes" – that is to say the material changes directly from a solid into gas.)

Mars is a difficult object to study with a small telescope, even when near oppo-sition, because a reasonably high magnification is needed. A 3-inch refractor will show the polar caps and the main dark areas, but do not expect too much unless your telescope has an aperture of at least 6 inches. For a drawing, a scale of 2 inches to the planet's diameter is customary; when the phase is evident, the disk should always be drawn to the correct shape. The map given in the Appendix 5 (page 175) shows the features I have been able to see with a 12½-inch reflector. (I have retained the older names; they have been revised in view of the space-craft results so that, for example Mare Acidalium has become Acidalia Planitia.) Much the most promi-nent of the dark features is the Syrtis Major, now known to be a high plateau from which the dusty layer has been blown away, exposing the darker rock beneath (Fig. 9.5).

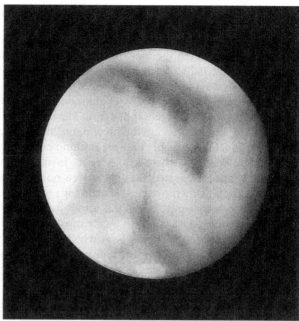

Fig. 9.5. Mars. Note the V-shaped Syrtis Major. (Paul Doherty, 12 April 1982, 2150 GMT, 419 mm reflector, ×248.)

At the start of an observing session, begin by looking at Mars for some time until your eye has become thoroughly dark-adapted. Then sketch in the main details, using a moderate power; next, change to a higher power and add the more delicate detail. As soon as this has been done, put the data in your sketch: time (GMT), power seeing conditions and any exceptional circumstances. Mars spins on its axis in 24 hours 37 minutes, so that the drift of the markings across the disk becomes noticeable over even a short period. Obviously, any particular marking will pass the central meridian about half an hour later each night; tables given in publication such as the *Handbook of the British Astronomical Association* make it easy to work out the longitude of the central meridian for any particular moment.

Owners of larger instruments may care to look for the two tiny moons, Phobos and Deimos. Both are veritable dwarfs less than a dozen miles in diameter, so that even when Mars is near opposition they are difficult to glimpse. I have seen them both with a 15-inch reflector, and keener-eyed observers should catch sight of them with a 12-inch when conditions are first-class.

A rather stupid mistake on my part may serve to show that it is not wise to reject an observation because it does not "fit in" with what is expected. I was once observing Mars with my 12½-inch reflector, when I recorded a minute starlike point, clearly visible only when Mars itself was hidden by an occulting bar, which I took to be Phobos. I then consulted my tables, and found that Phobos was not in fact anywhere near the position recorded. I therefore dismissed the observation, as either a mistake or else an observation of a faint star. It was only on the following day that I found that the observation itself was perfectly correct; I had made a slip in my calculations.

Phobos is a peculiar little body. It whirls round Mars at a distance of only 3,800 miles above the surface, about as far as from London to Aden, and it completes one revolution in only $7\frac{1}{2}$ hours. So far as Phobos is concerned, the "month" is shorter than the "day", and to a Martian observer Phobos would seem to rise in the west, gallop across the sky – taking only $4\frac{1}{2}$ hours to pass from horizon to horizon – and set in the east. Neither it nor Deimos would be of much use as a source of moonlight, and Deimos would indeed look like a large, dim star.

Both satellites were photographed from Mariner 9 and the Vikings. Each is irregular in shape, and each is pitted with craters. Phobos and Deimos are quite unlike our own Moon, and probably they are nothing more than ex-asteroids which were captured by Mars in the remote past, Iosif Shklovsky, a famous Russian astronomer, once suggested that they were nothing more nor less than hollow space-stations, launched by the Martians for reasons of their own; but I fear that the latest probes have put paid to this attractive, if somewhat remarkable, idea!

To many people, Mars seems to be the most intriguing world in the entire Solar System. If all goes well, it should be reached by men before the end of the present century. Whether any trace of life will be found is a question which ought to be solved within the next few decades; at present, the jury is still out.

Chapter 10

The Outer Planets

As soon as we look at a scale map of the Solar System, it is seen that the division of the planets into two main groups is very pronounced. Between the orbits of Mars and Jupiter there is a wide gulf of more than 300 million miles.

Nearly 200 years ago, Johann Bode suggested that there might be a small planet revolving round the Sun at a distance of about 260 million miles. There were sound reasons for believing that he might be right, and towards the end of the century a group of six leading astronomers, headed by Schröter and the Baron von Zach, began a systematic search for the missing body. Oddly enough, they were forestalled. Before the scheme was in working order, Giuseppe Piazzi at Palermo happened upon a star-like object which turned out to be a small world circling the Sun at almost the correct distance. It was named Ceres, in honour of the patron goddess of Sicily.

The trouble about Ceres was its size – or, rather, lack of it. The diameter turned out to be only 600 miles, too small for it to be regarded as a proper planet. The "Celestial Police" were not satisfied, and continued to hunt. Between 1801 and 1808 they found three more small bodies: Pallas, Juno and Vesta, and it was thought that these might be only the senior members of a whole swarm. However, no more seemed to be forthcoming, and the "Police" gave up. It was 1845 before a fifth "asteroid" (Astraea) was discovered, but since 1848 no year has passed without many new discoveries, and by now the grand total of known asteroids has passed 90,000. No doubt tens of thousands more exist.

Ceres remains the largest of these "Main Belt" asteroids, and of the rest only Pallas and Vesta have diameters of over 300 miles: Vesta alone is visible with the naked eye – just. Most are true midgets; for example one named asteroid, Hathor (No. 2340) has a diameter of about 22 yards, much the same as the length of a cricket pitch. With an ordinary telescope, even the largest asteroids look exactly like faint stars.

There are dozens of asteroids which can exceed the twelfth magnitude, and hunting and photographing them is a pleasant pastime; I once spent an evening searching for known asteroids with a 6-inch refractor, and observed fifteen of them in only two hours, though I could not identify many of them until I re-observed them later. The procedure is to look up the position of a suitable asteroid, using an almanac or the BAA *Handbook*, and plot it on your star chart. then go to the telescope and search until you have found the correct star-fields. The asteroid will look stellar, but when you look again, a night or two later, the stars will look the same –

Fig. 10.1. Apparent shift of the minor planet Pallas over a period of 24 hours. (Patrick Moore, 3-in. refractor.)

while the asteroid will have shifted as shown in Fig. 10.1. Of course, photography makes the hunt much easier.

Not all asteroidal bodies keep to the Main Belt between the paths of Mars and Jupiter. Some swing inward, and may come close to the Earth; we are not immune from collision, and it is true that if we were hit by a body of, say, a quarter of a mile across, the damage would be colossal. (According to one plausible theory, this did happen around 65 million years ago, and the resulting change in climate wiped out the dinosaurs.) These "Earth-grazers" are much more frequent than used to be thought, and if we are stuck we can only hope that we will manage better than the dinosaurs evidently did. There are a few asteroids which move in paths which take them very close to the Sun, so that at perihelion they must be red-hot; others move well beyond the Main Belt, and there are some – the 'Trojans' – which move in the same orbit as Jupiter, though they keep prudently either well ahead of or else well behind the Giant Planet and are in no danger of being swallowed up. Yet others inhabit the far reaches of the Solar System. Chiron, the first of these to be found (in 1977) spends most of its time between the orbits of Saturn and Uranus.

All these exceptional asteroids are very dim, either because they are extremely small or because they are a very long way away. A few can occasionally come within the range of a moderate telescope, but most do not.

In a book written by the well-known popularizer G. F. Chambers, published in 1898*, it is stated that "the asteroids are of no interest to the amateur observer, in fact they are of no real interest to anybody". The first of these statements may be true for the owner of a telescope such as a 3-inch refractor, but the second is not. Asteroids have been surveyed by space-craft, and have been found to be fascinating objects; some are of bizarre form (one, 216 Kleopatra, looks uncannily like a dog's bone!). A space-craft has even made a controlled landing on an asteroid, 133 Eros, which can occasionally pass Earth to a distance of only about 16,000 miles. But asteroid studies are in the main beyond the scope of this book, so let us pass on.

---

*Incidentally, this was the first book on astronomy that I ever read, and I still have it, but I hasten to add that I did not read it in 1898; I did so in 1929!

Beyond the main asteroid zone we come to mighty Jupiter, giant of the Solar System. Though it never approaches us much within a distance of 360 million miles, well over a thousand times as remote as the Moon, Jupiter still shines so brilliantly in our skies that it cannot be mistaken for a star. It is outshone only by Venus and, very occasionally, by Mars.

Jupiter's vast globe could contain 1,300 bodies the size of the Earth, but it is not so massive as might be supposed. If we could put Jupiter in one pan of a pair of scales, we would need only 318 Earths to balance it. This must mean that Jupiter is less dense than the Earth, and the density works out at only 1.3 times that of water.

Jupiter is not a rocky body like the Earth or Mars. When we look at it through a telescope, what we see is not a hard surface, but a cloudy vista with details which change not only from night to night, but from hour to hour. We must not, however, draw any comparison with Venus. Jupiter's "atmosphere", to use the word in a broad sense, is so deep that it merges with the true "body" of the planet.

It used to be thought that Jupiter consisted of a rocky core, overlaid by a 15,000-mile thick layer of ice which was again overlaid by the atmosphere. Modern ideas are very different. There is certainly a silicate core at a temperature of at least 20,000 °C, above which come layers of hydrogen in various states – liquid metallic hydrogen round the core, liquid molecular hydrogen next. The gaseous atmosphere is around 6,000 miles deep; the main constituent is hydrogen, while most of the rest is made up of helium, and there are hydrogen compounds such as ammonia and methane (it has been said that Jupiter must be a rather smelly world as well as a poisonous one!). Since the visible surface is gaseous, it is always changing. Moreover the rapid rotation means that the features are carried across the disk so quickly that the shift is obvious after only a minute or two.

In a telescope Jupiter appears as a yellowish disk, flattened at the poles and crossed by dark streaks known as belts. Closer inspection shows up features such as brightish patches, wisps and spots, all of which are phenomena of the high atmosphere. The belts, due to droplets of liquid ammonia, dominate the scene; usually there are two main belts, one to either side of the Jovian equator, with others at higher latitudes. They vary in appearance, and I remember one period (1962-64) when the equatorial belts merged to form a "wedge". On the classical model, gases warmed by the internal heat of the planet rise into the upper atmosphere and cool, producing clouds of ammonia crystals floating in the gaseous hydrogen; these clouds form the bright zones, which are higher and colder than the dark belts.

Spots can become prominent, but usually for only a limited period. The chief exception is the Great Red Spot, which has been seen on and off (more on than off) ever since the 17th century, and at its best is brick-red, though lately the colour has been less pronounced (en passant, we are not quite sure what causes the redness; phosphorus has been suggested). It can be 22,000 miles long and 7000 miles wide (Fig. 10.2, overleaf), and though it sometimes disappears for a while it always returns. Its longevity is presumably due to its great size; we now know it to be a whirling storm – a phenomenon of Jovian meteorology. Between 1901 and 1941 there was a feature in the southern hemisphere (the South Tropical Disturbance) which moved in the same latitude as the Red Spot, but had a shorter rotation period, so that from time to time it caught up with the Red Spot and "lapped" it

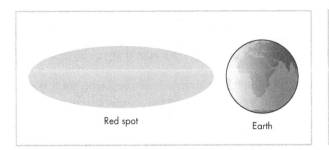

Red spot

Earth

Fig. 10.2. Size of the Great Red Spot.

with fascinating interactions. Nothing quite like it has appeared since, but there was a remarkable event in 1994, when a comet known as Shoemaker-Levy (in honour of its discoverers) crashed into Jupiter, producing huge scars which persisted for months and were visible with a very small telescope. If that comet had impacted Earth, the results would have been dire!

Several space-craft have encountered Jupiter, notably the Pioneers, the Voyagers and, most recently, Galileo, which was launched in 1989 and orbited the planet from 1995 to 2003 when it plunged in to the Jovian atmosphere and was destroyed. Most of what we know about Jupiter has come from these probes, so is there any point in continuing with telescopic observation from Earth? Again the answer is 'yes', because Jupiter is always changing, and one never knows what will happen next (Fig. 10.3).

Fig. 10.3. Photograph of Jupiter, showing the belts and the Great Red Spot. (Photographed by Russell Hawker in February 2003 with a 12-inch reflector.)

Jupiter's quick rotation means that one cannot afford to linger when making a disk drawing. The sketch should be completed in less than 10 minutes, as otherwise the drift of the surface features will introduce errors. As in the case of Mars, the main details should be filled in first; the time should then be noted, after which the magnification can be increased and the finer details added.

One minor irritation is that one cannot use a pencil compass to draw the outline of the disk. The polar compression amounts to 6,000 miles (as against 26 miles in the case of the Earth) so that it cannot be neglected, and shaping the outlines free-hand is a tedious process. I have found that the best solution is to obtain a stock of blanks, as shown in Fig. 10.4. These blanks are not expensive to have printed, and any local firm will make them at low cost. They can of course be photocopied.

Rotation periods of special features are best determined by the method of transits. There is no analogy with the solar transits of Mercury and Venus, and the word is used to denote the time when the feature under study passes across the central meridian of Jupiter.

What is done is to estimate the time of transit to the nearest minute, which is quite adequate. A measuring device is naturally helpful, but visual estimates can be made quite accurate enough for most purposes, and Jupiter rotates so rapidly that it is often possible to time 20 or 30 transits per hour. There is a standard nomen-clature, and this is given in Appendix VII, together with an extract from my own notebook that may prove helpful. It is hardly necessary to add that a reliable watch is essential – and make sure that it is set to the correct G.M.T.!

Once the transit has been found, the longitude of the feature can be found by means of the tables in the B.A.A. *Handbook*. This is an easy process, and involves nothing more frightening than simple addition.

It would be idle to pretend that visual observations of Jupiter's surface are as important as they used to be before the space-probes flew, but they are still well

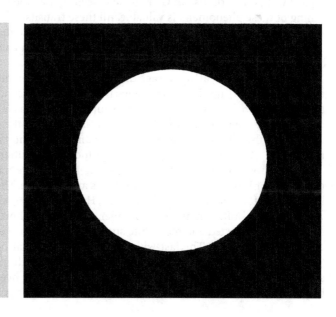

Fig. 10.4. This diagram can be scanned or photocopied so that you can run off your own stock of blanks.

worth making, and an instrument such as a 6-inch reflector can give good results. I remember that once, when Jupiter was well north of the celestial equator, I was able to watch a complete rotation, and secured two transit observations of the Red Spot on the same night.

Jupiter has an extensive satellite family. The principal four were studied in 1610 by Galileo, with his primitive telescope, and are known collectively as the Galileans; their individual names are Io, Europa, Ganymede and Callisto. All four are large; Europa is slightly smaller than our Moon, Io slightly larger, and Ganymede and Callisto much larger – indeed Ganymede is slightly larger than Mercury, though less massive. All the other satellites are tiny, and beyond the range of most amateur telescopes; some of them have retrograde motion, and are presumably captured asteroids.

It has been said that "there is no such thing as an uninteresting Galilean". This is certainly true. Space-craft results have shown that Ganymede and Callisto are icy and cratered and Europa icy and smooth, while Io is wildly volcanic, with eruptions going on all the time. It is thought that oceans of ordinary water lie below the icy surfaces of Europa and Callisto – so can there be any form of life? It sounds improbable, but we cannot be sure. One thing is definite: astronauts can never visit Io. Not only is it unstable, with sulphur volcanoes, but it moves within the strong belts of radiation known to surround Jupiter. Indeed, Io must be just about the most lethal world in the entire Solar System.

Surface details on the Galileans are quite beyond the range of small or moderate telescopes, but at least the movements of the satellites can be studied, and are fascinating to follow from night to night. Since all four revolve approximately in the plane of Jupiter's equator, they appear to keep in almost a straight line, but it often happens that one or more of them is missing. A satellite may pass in front of Jupiter, appearing in transit; it may pass behind, and be occulted; it may pass into Jupiter's shadow, suffering eclipse. The transits are particularly striking. In Plate VIII(c), a typical view, the dark disk of Ganymede is seen against the Jovian clouds. Accurate timing of these phenomena is valuable. All these transits, eclipses and occultations are predicted for each year in the B.A.A. *Handbook*, and in many almanacs.

Jupiter does have a ring system, but the rings are thin and dark; they are quite beyond the range of ordinary telescopes.

Far beyond Jupiter, at an average distance of 886 million miles from the Sun and a minimum of 741 million from the Earth, lies Saturn, second of the giant planets. In itself Saturn is less important than Jupiter; it is smaller, with an equatorial diameter of 75,000 miles and a mass of 95 times that of the Earth, and it is made up in much the same way. It is even colder than Jupiter, since the temperature never rises above –240 degrees Fahrenheit, and it too must be utterly lifeless.

Saturn shows belts and spots, but surface features are much less conspicuous than those of Jupiter, and well-marked spots are very rare. The last really spectacular outbreak took place in 1933, when W. T. Hay (Will Hay), a famous comedy actor who was also a skilled amateur astronomer, discovered a short-lived white spot near the equator. I detected a fainter white spot in 1962, but it never become prominent, and soon faded away. The last white bright spot was that of 1990, discovered by S. Wilber in much the same latitude as the earlier ones; within a few days it had been spread out by Saturn's strong equatorial winds, and had been transformed into a bright zone which extended all round the equator and persisted for some weeks.

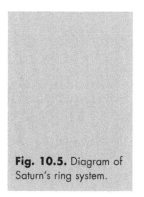
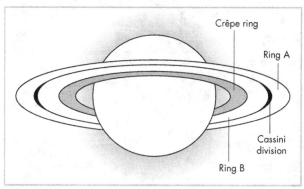

Fig. 10.5. Diagram of Saturn's ring system.

In many ways Saturn is not unlike Jupiter, but it is less changeable; it too seems to have a silicone core, overlaid by liquid hydrogen and then the cloudy "atmosphere"; the overall density of the globe is actually less than that of water. Most of our detailed knowledge has come from the two Voyager probes, which bypassed Saturn in 1980-81.

Of course the chief glory of Saturn is in its ring system. Christiaan Huygens, the leading telescopic observer of the 17th century, was the first to interpret them, and described "a flat ring", which is inclined to the ecliptic and which nowhere touches the body of the planet". Actually there are three main rings, two bright and one dusky (Fig. 10.5). The whole system has a diameter of almost 170,000 miles.

The rings are made up of countless icy particles, all whirling round Saturn in the manner of dwarf moons; any solid or liquid ring would be promptly disrupted by Saturn's powerful pull of gravity. They may be due to material 'left over', so to speak, when Saturn was formed; alternatively they could be the débris of a former icy satellite which was broken up. Pictures from the space-craft show that they are very complex, and there are also some rings beyond the main system which are very elusive indeed.

A 3-inch telescope will show the rings, but in a 6-inch the sight is glorious indeed, and Saturn is without doubt the most superb object in the heavens. It is unique in its glory, and it is a sight never to be forgotten.

Details can be seen in the ring-system. The bright rings, A and B, are separated by a dark area known as Cassini's Division, in honour of its discoverer. The Division is a true gap, and is due to the disturbing influences of Saturn's satellites. There is a second gap in Ring A (Encke's Division) which is not difficult to see when the ring system is well displayed.

Though the rings cover so wide an area, they are strangely thin and their thickness is less than a mile, so that when they are lying edgewise-on to the Earth they become hard to see. The drawings in Fig. 10.6 show what happens. The rings were edge-on in 1995, and will be so again in 2008. On such occasions small telescopes will lose track of the rings for a while, though I have always been able to follow them throughout with any telescope of over 8 inches aperture.

Saturn is an awkward object to draw, but there is no "short cut", as in the case of Jupiter. Stencils can be made to allow for the polar flattening of the disk, but the

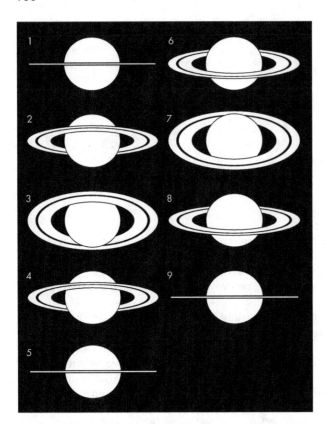

Fig. 10.6. Aspects of Saturn's rings. One full cycle is shown; the rings are closed in position 1, 5, and 9; the southern face of the ring is shown in 2–4, the northern face in 6–8.

rings have to be sketched freehand. Unfortunately it is not possible to prepare one standard drawing and use it as an outline for weeks on end, as the presentation of the rings alters perceptibly even over short periods. Always note the positions of the belts, and also the shadow effects – rings on disk, disk on rings; and watch out for anything unusual, such as a new white spot.

Saturn has a wealth of satellites, but only Titan is large. Almost any telescope will show it. Next in order of brightness come Iapetus and Rhea, which can be seen with a 3-inch refractor; Dione and Tethys, visible with a 4-inch and perhaps with a smaller instrument; and then Enceladus, Mimas and Hyperion, which are much dimmer. To see the smaller satellites a large telescope is needed. Many have been found – the total is now 30 – but most of the very small satellites are no doubt ex-asteroids. This may also be true of Phoebe, discovered as long ago as 1898, which is a very long way from Saturn and has retrograde motion. I have just glimpsed it with my 15-inch reflector.

It is interesting to estimate the magnitudes of the satellites. Field stars can be used when available – but make sure to identify both stars and satellites correctly; a stupid mistake of my own shows what is meant. In 1966 the rings were edge-on, and Audouin Dollfus, in France, detected a very small inner satellite (now called Janus) which is normally very hard to see. Looking at my observations, made with a 10-inch refractor, I found that I had seen Janus twice without realizing that it was new! Iapetus is of special interest; when west of Saturn it is easy to see, but when east of Saturn it becomes much fainter. We now know that this is because its two hemispheres are very unequal in reflecting power – one icy, the other as black as a blackboard.

All the main satellites have been surveyed by the Voyager probes. Most are icy and crater-scarred, but Titan is the exception; it is over 3000 miles in diameter – bigger than Mercury – and has a dense atmosphere, so that even the Voyagers could not see through to the true surface. Nitrogen makes up the bulk of the atmosphere, but most of the rest is methane, and the surface temperature is very low. Life of any kind there seems highly improbable, but we may know more in late 2004, when the Huygens space-craft, taken there by the Cassini probe, makes a controlled landing on the surface. Whether it will come down on an ice-sheet, or splash into a chemical ocean, remains to be seen.

Saturn was the outermost of the planets know in pre-telescope times. In 1781 William Herschel discovered Uranus, which is on the fringe of naked-eye visibility and is easy to see with binoculars. It is never less than 1,600 million miles from the Earth; it qualifies as a giant, with a diameter of over 30,000 miles, but it is much smaller than Jupiter or Saturn, and its composition is quite different. While Jupiter and Saturn are well described as gas-giants, Uranus and Neptune are probably better described as ice-giants. Uranus is made up largely of "ices", but these need not be in solid form; we have a mixture of water, methane and ammonia, with traces of other substances plus a certain amount of "rock", but inside planets such as Uranus the conditions make all these materials behave as liquids. Uranus, also among the four giants, has little or no source of internal heat. The outer atmosphere is made up chiefly of hydrogen, with a good deal of helium, and also some methane; and this absorbs red light, which is why Uranus appears greenish.

In some way, Uranus is a celestial oddity. Whereas most of the planets have their axes of rotation inclined to the perpendicular to their orbital planes by 20 to 30 degrees (Fig. 10.7), Uranus has an inclination of more than a right angle. Consequently, the "seasons" must be peculiar, particularly as the Uranian "year" is 84 times as long as ours. First part of the northern hemisphere, then part of the southern, is plunged into darkness for 21 years, with a corresponding period of daylight in the opposite hemisphere. Sometimes a pole appears in the middle of the disk, as in 1985, while at other times the equator is presented. Uranus is a quick spinner – its "day" is a mere $17\frac{1}{2}$ hours – and its rotation is technically retrograde, though not generally regarded as such.

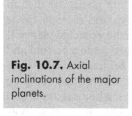

Fig. 10.7. Axial inclinations of the major planets.

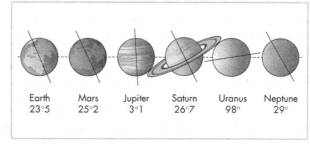

Earth	Mars	Jupiter	Saturn	Uranus	Neptune
23°5	25°2	3°1	26°7	98°	29°

One space-craft has bypassed Uranus; Voyager 2, in 1986. There were none of the vivid hues or striking clouds of Jupiter and Saturn, and Uranus appeared to be a rather bland world, though some cloud structures were seen, and were later recorded also by the Hubble Space Telescope. There is a system of thin, dark rings, well beyond the range of amateur telescopes, and indeed there is little to be seen on Uranus even with high magnification. It is interesting to estimate the planet's apparent brightness, because there seem to be variations which may be linked with disturbances on the disk. The method is to compare Uranus with a nearby star of known magnitude, just as is done for a variable star (see Chapter 15), but there is the complication that unless a very low power is used, Uranus will show a perceptible disk. I remember that in 1955, when Uranus and Jupiter lay close together in the sky, I tried to compare Uranus and Ganymede and Callisto, but the planet was obviously larger than the satellites and had a lower surface brightness, so that my estimates did not mean a great deal.

Uranus has an extensive satellite family, but only five were known before Voyager 2 pass. Of these Titania and Oberon can be seen with a telescope of around 10 inches aperture; Aeriel and Umbriel are much fainter, and I doubt whether Miranda can be glimpsed with anything less than a 24-inch. Titania, the largest member of the family, is only 980 miles in diameter.

Neptune, last of the giants, is often referred to as the twin of Uranus, but apart from size and mass the two are not identical. Neptune does have a strong internal heat-source, and it does not share Uranus' exceptional tilt. It was discovered in 1846 by the German astronomers Johann Galle and Heinrich D'Arrest, not by accident; Uranus had not been moving quite as expected, so that it was being pulled out of position by a more distant planet. The French astronomer Urbain Le Verrier worked out where the planet ought to be – and he was right. Neptune can be seen with binoculars, but in a telescope of less than around 10 inches aperture it looks stellar, and not even large instruments will show definite detail on its tiny, bluish disk. Most of what we know about it comes from observations by Voyager 2, which bypassed it in 1989, and, more recently, from the Hubble Space Telescope. Neptune is a much more dynamic world than Uranus, and it moves slowly against the starry background, so that once it has been identified telescopically it is not hard to find again. Its senior satellite, Triton, is larger than any of the satellites of Uranus, and I have seen it with my $12\frac{1}{2}$-inch reflector; observers with keener eyesight than mine tell me that it can be glimpsed with a 6-inch. The other Neptunian satellites are beyond the range of amateur-owned telescopes.

With the discovery of Neptune, the Solar System seemed to be complete, but the movements of the outer planets were still not in perfect accord with calculation, and Percival Lowell – best known for his admittedly wild theories about life on Mars – set to work in order to find the position of yet another planet. He died in 1916, but the search continued at his Flagstaff observatory, and in 1930 Clyde Tombaugh located a dim, starlike object which proved to be the expected planet. It was named Pluto, after the God of the Underworld, and indeed it must be cold and lonely, with the Sun looking like only a tiny though intense brilliant disk.

It all seemed to be the most satisfactory – but it was not. Pluto turned out to be smaller than the Moon, with a diameter of a mere 1444 miles; a body as small and lightweight as this could not noticeably perturb the movements of giants such as Uranus and Neptune – and yet it was by these very perturbations that Lowell had calculated its position. Was his success sheer luck, or could there be another planet out there? And was Pluto really worthy of true planetary status? Its orbit is unlike that of any other planet. The orbital period is 247.7 years, and the distance from the Sun ranges from 2,766 million miles at perihelion out to as much as 4,566 million miles at aphelion. At its closest to the Sun, therefore, Pluto is actually closer-in than Neptune, but the orbit is tilted at an angle of 17 degrees, so that there is no danger of a collision. Pluto passed the perihelion in 1989, and reached magnitude 14, so that a modest telescope would show it. It can be identified in the same way as an asteroid, though with more difficulty because it moves so slowly. Figure 10.8 shows the orbit. It does have a satellite Charon, half the size of Pluto itself, but no amateur telescope has a hope of showing it. It moves round Pluto in a period of 6 days 9 hours, and this is also Pluto's rotation period so that to an observer on Pluto, Charon would be motionless in the sky.

No large planet has been found beyond Pluto, but during the past few years large numbers of asteroid-sized bodies have been detected, moving in orbits close to and beyond that of Pluto. They make up what is called the Kuiper Belt in honour of the Dutch astronomer Gerard Kuiper, who suggested its existence. Some KBOs (Kuiper Best Objects) are of considerable size, and one, Quaoar, may be 1000 miles across – not so very different from Pluto. It now seems that we really have to demote Pluto, and re-classify it as a KBO, but at least it is the brightest member of the swarm. Of the rest, even Quaoar is beyond amateur range.

This survey of the planetary system has been very incomplete, but all I have been tried to do is to indicate the main points of interest to the modestly-equipped amateur. There is plenty to see, and the planets are fascinating companions.

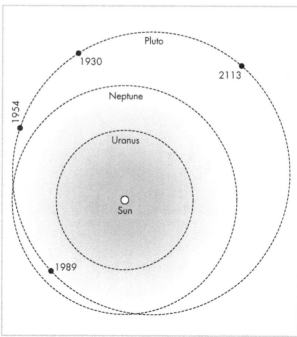

Fig. 10.8. The orbit of Pluto. Perihelion, 1989; aphelion, 2113.

Chapter 11

Comets and Meteors

A brilliant comet, with a tail stretching right across the sky, must be one of Nature's grandest spectacles. Small wonder that comets caused fear and panic in ancient times, when they were believed to be heralds of disaster. Shakespeare wrote in *Julius Caesar*:

> When beggars die, there are no comets seen
> The heavens themselves blaze forth the death of princes,

and the feeling is not entirely dead even today.

A comet is quite unlike a planet, and has been described as a "dirty ice-ball". The only reasonably substantial part is the nucleus, made up of rocky fragments held together by ices, so that even a large comet is of very slight mass. True, if we were hit by a cometary nucleus the damage would be tremendous – recall that a comet actually did strike Jupiter in 1994, so that the Earth is not immune. Fortunately, most comets keep at a respectful distance.

One point is worth clearing up. If you see something moving across the sky against the starry background, it cannot be a comet, for the simple reason that a comet is millions of miles away, and has to be watched for hours before any individual motion becomes noticeable. A moving object has to be much closer to the ground so that unless it is a distant aircraft or an orbiting artificial satellite it will almost certainly be a meteor – a tiny particle dashing into the upper air and burning away to produce a shooting-star.

Comets move round the Sun, but in most cases (not all) their orbits are very eccentric, and the revolution periods may be long. Only one bright comet (Halley's) has an orbital period of less than a century. There are many comets with much shorter periods, but few of these ever become bright enough to be seen with the naked eye.

Most people think that a comet should look like a bright fuzzy mass, with a long tail streaming out. Great comets do indeed appear like this, but really brilliant visitors have been rare of late, and most comets are much less imposing. I remember showing a telescopic comet to a friend of mine who knew little about astronomy and cared less. His comment was that the comet looked like "a small lump of cotton-wool", and in all honesty I could not disagree with him!

A comet is made up essentially of three parts: a nucleus, a head (coma) and perhaps a tail or tails. The nucleus – the "dirty ice-ball" – is never large, and even

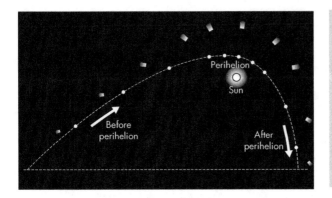

Fig. 11.1. Direction of a comet's tail with respect to the Sun.

in a major comet the diameter of the nucleus is seldom as much as 30 miles. The nucleus of Halley's Comet, surveyed by space-craft in 1986, was shaped rather like a peanut, with a longest diameter of just over 9 miles.

When a comet moves in toward perihelion, and is warmed, the ices in the nucleus begin to evaporate. Gas flows outward, to form the coma. A coma may be huge (that of the Great Comet of 1811 was larger than the Sun), but its gases are immensely rarefied, and correspond to what a laboratory technician would call a satisfactory vacuum.

Tails are of two main types: gas (ion) and dust. The Sun sends out streams of particles all the time, producing the solar wind; these particles striking a comet drive out enough gas to produce a straight tail, often of tremendous length. Light also can exert a pressure, and this drives out dust-particles from the comet to form a tail which is usually curved. Tails develop only when a comet moves in toward perihelion, and disappear when the comet moves back into the remote cold parts of the Solar System. Many small comets fail to develop tails of either type. Note that a tail always points more or less away from the Sun (Fig. 11.1), so that after perihelion, when the comet is moving outward, it travels tails first.

A comet loses material every time it forms a tail, and it is bound to waste away, so that by now all the short-period comets are very dim. In general they are named after their discoverers (occasionally, after the mathematician who first worked out the orbit), and in most cases their orbits are well known, so that we always know when and where to expect them. Some can be followed all the time, such is Encke's Comet, with a period of 3.3 years. Its orbit is shown in Fig. 11.2.

Many short-period comets have their aphelion points close to the orbit of Jupiter, and make up the "Jupiter family". Jupiter's powerful pull of gravity has marked effects, and a comet may have its path completely altered. Thus Lexell's Comet of a 1770, which became an eye object, made a close approach to Jupiter a few years later; its orbit was violently twisted, and we have no idea where the comet is now.

Special mention must be made of the most famous of all comets, Halley's. In 1682 a bright comet appeared, and was seen by Edmond Halley, who later became England's second Astronomer Royal. At that time most people (even Isaac Newton) believed that comets came from outer space, travelled in straight lines and then departed for ever, Halley was not so sure. He realized that the 1682 comet moved in the same way as comets seen in 1607 and 1531; could they be one and the same?

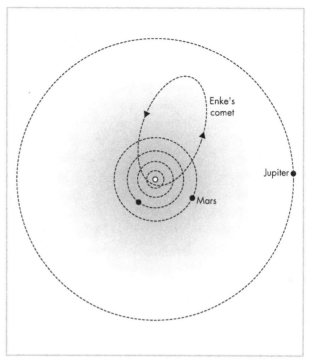

Fig. 11.2. Orbit of Encke's Comet.

If so, it would be back once more in 1758. Halley knew that he would not live to see it (he died in 1742) but on Christmas Night 1758 the comet was duly found, by a German amateur, where Halley had expected it. The comet passed perihelion in 1759, since it has returned in 1835, 1910 and 1986, and is due once more in 2061. We know where it is now, but it is too faint to be seen, and probably we will not pick it up again until after 2050.

There are many historical records of Halley's Comet. For instance at the return of 837 it was apparently bright enough to cast shadows, and it shone down in 1066 before the Battle of Hastings; in the celebrated Bayeux Tapestry it is shown clearly, with the Saxons looking on aghast and King Harold toppling from his throne. It was bright in 1910, but was badly placed in 1986, though it was an easy naked-eye object for some weeks. Five spacecraft were sent to it: two Japanese, two Russian and one European. The European probe, named Giotto in honour of the Florentine painter who showed it in a famous painting, penetrated the comet's coma and obtained close-range images of the nucleus before its camera was put out of action by the impact of a particle probably about the size of a grain of rice. Water ice was found to be the main constituent of the nucleus, but there was a coating of dark material; jets were much in evidence. Some onlookers were disappointed with Halley's Comet, because it was not so conspicuous as it had been at most previous returns; sadly it will again be badly placed in 2061, but in 2139 it should be magnificent. Wait for it!

Some comets have orbits which are parabolic or hyperbolic (Fig. 11.5) so that after perihelion they will never return. It used to be thought that these comets

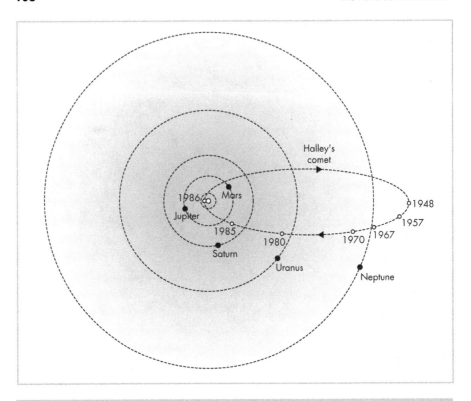

Fig. 11.3. Orbit of Halley's Comet. Aphelion, 1948; perihelion, 1986.

came from interstellar space, but now seems that they originate in what is called the Oort Cloud (after the Earth astronomer Jan Oort), at least a light-year from the Sun, where there are many millions of these icy bodies – cosmic débris. If an Oort object is perturbed for any reason, it may start to fall sunward, finally invading the inner Solar System and coming within reach of our telescopes. One of several things may then happen to it. It may fall into the Sun, and be destroyed; it may encounter a planet (usually Jupiter) and be forced into a short-period orbit; it may hit a planet, as Comet Shoemaker-Levy did in 1994, or it may simply return to the Oort Cloud. However, some comets are perturbed into open orbits, and have therefore paid their one and only visit to the Sun before being thrown out of the Solar System for ever.

Apart from Halley's all Great Comets have periods of so many centuries that they cannot be predicted. Several were seen during the 19th century, but between 1910 and 1996 there was a comparative dearth of them. There were, of course, many comets visible without telescopic aid, but faint objects at the limit of naked-eye visibility are very different from the spectacular Great Comets of the past. The Great Comet of 1843 had a tail which stretched right across the sky, while that of 1858 (Donati's) was peculiarly beautiful in view of its triple tail, curved like a scimitar. Another particularly brilliant comet was seen in 1882 (Fig. 11.6).

One of the most interesting comets of recent times was Arend-Roland, discovered in November 1956 by the two Belgian astronomers after whom it was named.

Fig. 11.4. Halley's Comet, with the Pleiades. (Photographed by the author.)

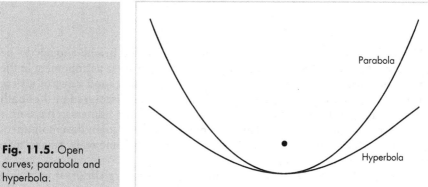

Fig. 11.5. Open curves; parabola and hyperbola.

Though hardly a "great comet" in the true sense of the world, it was a conspicuous naked-eye object for a short time in April 1957. On the 27th of that month, I had a particularly good view of it in the late evening; the nucleus lay close to the star Alpha Persei, and the long tail extended upwards from the horizon, so that it was a splendid sight in binoculars or a low-power telescope (to me, it seemed most impressive with binoculars). One of the interesting features of this comet was a curious "reverse tail". This "reverse tail" was still faintly visible in April 1957, while the nucleus of the comet then appeared almost stellar.

Fig. 11.6. Comet Linear, 2002 T7. (John Fletcher, Gloucester.)

There have been various comets which have become bright enough to be conspicuous. One was Ikeya-Seki, discovered by two Japanese astronomers in the summer of 1965. From Europe it was disappointing, though good views of it were obtained from the United States. Bennett's Comet of 1969, discovered by the South African amateur Jack Bennett (one of the most skilled comet-hunters in the world) did achieve prominence, and developed a long tail. But the main disappointment was Kohoutek's Comet of 1973, which was expected to become really spectacular, but which signally failed to do so.

When discovered, by Dr Lubos Kohoutek at the Hamburg Observatory, it was very faint, but it was also a very long way from the Sun, and early predictions indicated that towards the end of the year it might exceed magnitude – 10. Alas, it did not brighten up as had been hoped, and although it was visible with the naked eye it was by no means spectacular – certainly inferior to Bennett's Comet of four years earlier. However, it was interesting scientifically, and was found to be associated with a vast cloud of tenuous hydrogen; it was studied by the last crew of America's space-station Skylab, and a great deal was learned from it. The estimated period is 75,000 years, so that it will not come back in our time.

West Comet of 1975 was brighter, and almost qualified for the title of "great". Over Britain it made a brave showing in the dawn sky for several mornings in succession. During its flight round the Sun it showed obvious signs of disintegration, so that when it next returns – in many centuries from now – it will have none of its twentieth-century glory.

We had to wait until the century had almost run its course before we saw any more notable comets. The first was discovered in 1997 by a Japanese amateur, Yuji Hyakutake, and was really beautiful; it was decidedly green, and was brilliant for a few nights in the spring. Actually it was a small comet, but it came close to the Earth, and had a very long tail. It has a period of about 14,000 years, so that we have bade it farewell. It was outmatched in the following year by the comet discovered by two American observers, Alan Hale and Tom Bopp. This comet was superb; its head became brilliant, and there were two tails, one curved (dust) and one straight (gas), with indications of a third. It hung in the sky for months, and remained visible with the naked eye from July 1996 to October 1997 – easily a record, it must have been seen by more people than any other comet, and it is a great pity that it will not be back for 2360 years. Once it had departed, the night sky seemed strangely barren without it.

There is no knowing when the next Great Comet will appear. It may be this week; it may not be for decades. We can only hope.

It is a great pity that brilliant comets have been in such short supply lately. Indeed, the last really spectacular visitor was the "Daylight Comet" of 1910 (which is not periodical, and should not be confused with Halley's, which returned in the same year). Of course, there is no knowing when another great comet will appear. It could happen at any time.

When a short-period comet is due to return, its expected position at the time of anticipated recovery is given in a yearly publication such as the B.A.A. *Handbook*. The positions given are usually accurate enough for quick identification. Last time Encke's Comet came round, I picked it up without difficulty as soon as it came within range of my portable 3-inch refractor; and I am not, and never have been, a regular observer of comets. (True, Encke's Comet can be followed all round its orbit – but not with a 3-inch refractor!)

The chief scope from the amateur's point of view is that many comets appear unexpectedly, and completely "out of the blue". There is always the chance of making a discovery, and some amateurs are adept at it, so that they have established international reputations.

Comet-sweeping is therefore a worth-while occupation, but the beginner must resign himself to many disappointments and many hours of fruitless searching. He may not discover a comet for years, or he may never discover one at all. There is however a great consolation, since even if he fails to find a comet he will be certain to come across many stellar objects of real interest.

Good binoculars are very suited to comet-sweeping. George Alcock, one of the most successful of modern comet-hunters, never used a telescope – I do not believe that he ever owned one. He carried out his work with specially-mounted binoculars, and found four comets as well as several novae.

Telescopically, never use a high specification. What is needed is a wide field of view, and in any case a powerful eyepiece is of use upon a badly-defined, fuzzy object such as the average comet.

Having selected the region to be swept, the telescope is moved slowly along in a horizontal direction (if on an altazimuth mount), with the observer keeping a careful watch all the time. Stars, clusters and other objects will creep through the field, and the slightest relaxation of attention may mean that a vital comet is missed. At the end of the sweep the telescope is raised or lowered very slightly, and an overlapping sweep taken in the opposite direction. After this process has been carried on until the whole area has been covered, it should be repeated several times until the watcher is satisfied that no dim, misty object can have escaped him.

Much patience is called for, and things are made more difficult by the presence of star-clusters and nebulae, which look very much like comets. The name "star-cluster" speaks for itself, while a "nebula" is rather similar in appearance, and is made up either of stars or gas. If you are sweeping the heavens in search of a comet, and happen to find a misty object that is certainly not a star, it is unwise to jump to any conclusions. Reference to an atlas will probably show that the object is a cluster or a nebula that has been known for centuries*.

There is an interesting story about these clusters and nebulae. Charles Messier, a famous comet-hunter of the eighteenth century was persistently misled by unchartered stellar objects, and eventually he drew up a catalogue of "objects to avoid", rather as a navigator charts shoals in a strait. Nowadays Messier's comets are forgotten by all but a few enthusiasts, while his catalogue of clusters and nebulae remains the standard.

Many comets will lie somewhere near the Sun's line of sight when they are approaching perihelion, and nearly all remain undiscovered until they are well within the orbit of Mars, particularly as the average comet brightens up considerably as it draws near the Sun and the heat acts upon the particles in the nucleus. Consequently, the most promising areas of the sky for sweeping, for an observer in the northern hemisphere, are the west and north-west after sunset and the east and north-east before sunrise. It is also worth sweeping in the low north. It is no use beginning until the sky is really dark, since a faint comet will be drowned by any background light.

Though more new comets will be seen in these directions than in any others, there is no hard and fast rule. A comet may appear at any moment from any direction; it may have an open or closed orbit, it may be highly inclined, it may be moving in a retrograde or wrong-way direction. Comets have been called the stray members of the Solar System. Flimsy, harmless and of negligible mass, they are short-lived upon the astronomical timescale, and several short-period comets seen at several returns during the past have now vanished for good. Such are the comets of Biela and Brorsen.

Apart from comet sweeping, the amateur who has equipped himself with an equatorial mount, a measuring device and perhaps a camera, can do valuable work in checking the positions of comets from night to night. Mathematically-minded workers prefer to make a hobby of computing orbits. This is not an easy process, and real skill is needed, but anyone who has the necessary ability and patience will soon find that his services are in great demand – more especially if he is the owner of a computing machine!

The link between cometary and meteoric astronomy is perhaps shown most clearly by the interesting case of Biela's Comet, whose peculiar career caused many

---

*Odd things have happened now and again. Not long ago, one observer reported a "comet" that proved to be merely a reflection in his telescope, and there have been other similar cases. Even better was the "discovery" of a bright red star in the constellation of the Bull, not far from Aldebaran. It turned out to be Mars.

astronomers many sleepless nights. The comet was discovered by Biela, an Austrian astronomer, in 1826, and found to be identical with comets previously observed in 1772 and in 1805. It was one of Jupiter's short-period group, and had a period of about 6¾ years. It returned in 1832 as predicted, was missed in 1839 owing to its unfavourable position in the sky, and returned once more in 1845.

Up to then, the comet had behaved in a perfectly normal manner, but during the return of 1845-46 it astonished observers by splitting into two pieces. Where there had been one comet, twins could be seen, sometimes with a kind of filmy bridge between them. Sometimes the two were nearly equal, sometimes the original comet was the brighter. Both faded gradually into the distance, and the return of 1852 was eagerly awaited. This time the two comets were farther apart, the second following the first rather like a child following its mother. At the 1859 return conditions were again hopelessly bad, but in 1856–66 the comet should have been an easy telescopic object. Yet it was searched for in vain. There was no trace of it; so far as could be made out, Biela's Comet had disappeared from the Solar System. Comets have been nicknamed "ghosts of space", but no ghost could possibly have done a more successful vanishing act.

The next return should have taken place in 1872. Again the comet was absent, but in its place appeared a rich shower of meteors. Coincidence can be ruled out, and for years afterwards meteors were seen each year at the time when the Earth crossed the path of the dead comet. This shower is still active about November 28 annually, though it has now become very feeble.

It would be misleading to say simply that Biela's Comet "broke up" into meteors. There is more in it than this, and the position has been made clearer by the associations of other comets with other meteor showers; Halley's Comet, for instance, is linked with the meteor shower seen each year during the first week in May, and known as the Aquarids. Débris must be spread widely along the track of a comet, though once again it would be misleading to suppose that all meteors must be connected with comets.

Most people have wild ideas about the sizes of the particles which become incandescent and are rapidly burned up to become shooting-stars. Actually, the particles are very small. A body the size of a grape would produce a brilliant fireball, while the average bright meteor is due to a particle less than a tenth of an inch in diameter. Like comets, meteors are less important than they seem.

A meteor travels round the Sun in an elliptical orbit, sometimes as a member of a shoal ("shower meteor") or as a lone wolf ("sporadic meteor"). If it comes close to the Earth, and is moving in a suitable fashion, it may enter the upper atmosphere at a relative speed of up to 45 miles per second. Below an altitude of 120 miles or so, there is enough air to cause appreciable resistance; heat and visual radiation are generated, and the hapless meteor is destroyed, ending its journey in the form of fine dust. Millions of shooting-stars enter the Earth's atmosphere every day. Most are smaller than grains of sand; the so-called micro-meteorites, which have been investigated recently by means of sending high-altitude rockets above the densest layers of the atmosphere, seem to have diameters of something like 5/1000 of an inch, and may be similar to the particles which cause the glow of the Zodiacal Light.

Sporadic meteors may appear from anywhere at any time, but shower meteors are more obliging. If the Earth passes through an area in space which is rich in

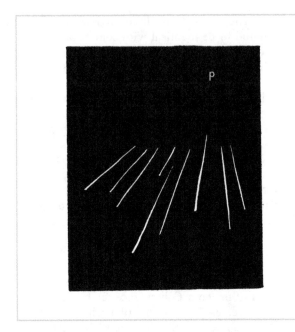

Fig. 11.7. Diagram to illustrate the principle of a meteor radiant. The meteors are assumed to be parallel, but to the observer the paths will seem to diverge from the point P.

meteors, the ordinary laws of perspective will cause the meteors to appear to radiate from one point. This is shown in Fig. 11.7, where all the meteors appear to converge towards a distant point P, which can be regarded as the apparent "radiant" of the meteors (Fig. 11.8).

A meteor shower is named according to the constellation in which the radiant seems to lie. For instance, one major shower visible each November has its radiant in Leo, the Lion, and is thus called the Leonid Shower; of course, this does not mean that all the meteors appear near Leo, but merely that if the paths were plotted back, they would converge to a small area in Leo known as the radiant. Similar, the October Orionids radiate from Orion, and the August Perseids from Perseus.

Some of the annual showers are more important than others, and a list is given in Appendix 11, but really spectacular displays are very rare. Such were the showers of 1833 and 1866, when the Leonids (associated with Tempel's periodical comet) were much more numerous than usual, and it was said that shooting-stars seemed to "rain down like snowflakes".

In fact, the Leonids had had a long and spectacular history, and had been consistent in providing major displays every 33 years. After 1866, the next was due in 1899 – but by then, unfortunately, the meteor swarm had been affected by planetary perturbations, and the main cluster missed the Earth, so that the expected display did not materialize. The next return was due in 1933 (not 1932), but again there was nothing of note.

Conditions seemed more promising for 1966. The Leonid displays of 1963 and 1964 showed an encouraging increase, and this was also true of 1965, though for that year the observations were hampered by the inconvenient presence of the full moon. Much was hoped for 1966, and earlier in the month I put out a television

Fig. 11.8. Illustrating a meteor radiant. The parallel lanes of the M6 motorway appear to diverge from a distant point when photographed from a bridge. (Photograph by the author.)

appeal for what is known officially as "audience participation". With me was H. B. Ridley, the Director of the Meteor Section of the British Astronomical Association. We announced that charts and 'answer cards' would be distributed, and during the next few days the B.B.C. dispatched more than 10,000 of these charts and cards to people who wrote in for them.

The result was a sad anti-climax. In Ireland, where I was observing, the skies were reasonably clear, but at an early stage it became evident that the Leonids were going to fail us yet again. We saw some meteors, and plotted a radiant, but the display was so poor that nobody would have noticed it except by careful, systematic watching. Matters were very different elsewhere. As seen from parts of the United States (Arizona, for instance) the hourly Leonid rate reached 100,000 – it was the greatest display of the century. Maximum occurred at about 12 hours G.M.T., while it was daylight in Europe; in fact, British observers missed the display by six hours. Yet the counts made by British amateurs were valuable scientifically.

The fact that the display was so brief proved that the meteors were "bunched" together, and were not spread all along the orbit of their parent comet in the same manner as the Perseids.

The Leonids are associated with Comet Tempel-Tuttle, which has a period of 33 years. The comet was due back in 1999, and we hoped for good displays. In fact there were rich but brief 'meteor storms' in November 1999, 2000 and 2001, but nothing to equal 1833 or 1866. I have to confess that I have had absolutely no luck with the Leonids; I have either been in the wrong place, or else clouded out. I doubt whether I will have another chance, unless I live to the advanced age of a hundred and ten.

To find out the speed, height and orbit of a meteor, three data must be provided: the point of appearance of the meteor, the point of disappearance, and the duration. Clearly it is necessary for the same meteor to be observed by two workers placed at least twenty miles apart (more if possible). A single observer cannot do much if he has to depend only upon his own labours.

No instruments are needed for meteor recording, but the observer has to have a really good knowledge of the constellations, as otherwise he will be unable to plot the track. The track must be plotted on a star map, but it is unwise to look down as soon as the meteor has vanished and try to record where it went, since errors are certain to creep in. The solution is to check the path by holding up a rod or stick along the track where the meteor passed, which will give you the chance to take stock of the background and ensure that no mistake has been made. When you are satisfied, either draw the path on your chart or note the exact positions of the beginning and end of the track, and then write down: time of start, duration, duration of luminous trail, brightness (compared with that of a known star or stars), colour (if any), and any special features.

Meteor watching is a lengthy and often a cold business. Standing out for hours during a January or February night is enough to chill the enthusiasm of the hardiest observer. Nevertheless, until recently all researches were based upon the patient work of amateurs, among whom the name of W. F. Denning will always be remembered. If it is fair to say "until recently", because in 1946 an entirely new method of recording was brought into operation, that of radar.

The passage of a meteor through the atmosphere has a pronounced effect upon the air-particles, and these effects can be detected by radar. Reduced to its barest terms, radar involves sending out an energy wave, and recording the echo as the wave is bounced back after hitting a solid object. A meteor trail is not of course a hard body, but it acts similarly, and radar detection of shooting stars has now been in progress for some time. The method is unhampered by clouds or daylight, and it would be idle to pretend that it has not affected the value of amateur visual work, though the naked-eye watcher can still make himself useful.

Casual meteors are fairly frequent, and a watchful observer will seldom fail to record fewer than five or six per hour, but it is of course far more entertaining (though not necessarily more useful) to concentrate upon some definite shower. Occasionally there will be so many meteors in quick succession that the watcher will be hard pressed to record them all, but this will not happen often, and there must be long periods of patient waiting.

It is interesting to note that visual meteors are twice as abundant in the period from midnight to 6 a.m. as during the period from 6 p.m. to midnight. In the evening, we are on the "rear" of the Earth as it moves in its orbit, so that visible meteors have to catch us up; in the morning hours we are in the "front" position, so that meteors meet us coming. More meteors are to be expected after midnight than before it, and obviously the morning meteors will have greater relative speed, just as a car moving at 30 m.p.h. and meeting a second car moving at 35 m.p.h. will be badly damaged if the collision is head-on, but only bumped if rammed from behind. It is the relative speed of a meteor which is the main factor in its brightness, so that the morning meteors will be more brilliant and hence easier to record.

Now and then the Earth is struck by a larger body, which can survive the complete drop to the ground and is then known as a meteorite. Most museums have collections of them and the expert can soon tell what is meteoritic material and what is not, though the layman can easily be misled. In general meteorites are divided into two classes, stones (aerolites) and irons (siderites), though in many cases both types of material are found.

It is important to remember that there is absolutely no connection between a meteorite and a shooting star meteor. Meteors are cometary débris; meteorites come from the asteroid belt, and it is fair to say that there is no distinction between a large meteorite and a small asteroid, though the term "meteorite" is restricted to objects which have actually landed on the Earth.

Large meteorites are rare, though some are colossal; the holder of the heavy-weight record is still lying where it fell, in prehistoric times, at Hoba West in South Africa. Nobody is likely to run away with it, since it weighs over 60 tons. Meteorite craters are found; one, in Arizona, in three-quarters of a mile across, and was certainly formed by an impact around 50,000 years ago. (Go there if you can. It is a well-known tourist attraction, not far from the town of Winslow; Svante

Fig. 11.9. The Meteor Crater in Arizona. (Photographed from the air by Efren Lopez.) (It really should be called the meteorite crater.)

Arrhenius, the famous Swedish scientist, once called it "the most interesting place on earth".) A major impact would be disastrous, but the Earth is a relatively small target in a vast expanse of space! (Fig. 11.9)

An object of some kind hit Siberia in 1908, blowing pine-trees flat over a wide area; nobody is sure whether it has to be classed as a meteorite or as the nucleus of a dead comet. Siberia was again hit in 1948, near Vladivostok, and this was certainly a meteorite, which broke up during descent and produced numerous small craters. Two interesting meteorites fell in Britain during the past half-century; one of these landed on Christmas Eve 1969. A meteorite flashed across the Midlands, attracting considerable attention, and then broke up, showering fragments in and around the village of Barwell in Leicestershire. The original weight of the meteorite must have been around 200 pounds. One fragment went through the window of a house in the village, and was later found nestling coyly in a vase of artificial flowers. There was also the Bovedy Meteorite, which shot across England and Wales and dropped fragments in Northern Ireland, causing the usual crop of flying reports. I missed it by two minutes. I had been in my observatory in Selsey, checking on variable stars, and had just gone indoors to change my charts when the meteorite passed over....

It is fitting to end this brief survey of the Solar System with the meteors and meteorites, its most insignificant members. We have described the Sun, the Moon, the planets and their satellites, the vivid glow of the aurora and the pale radiance of the Zodiacal Light, and the flimsy and unpredictable comets, so that there is a variety in plenty; but even the Sun itself is a very junior member of the Galaxy, and we must keep our sense of proportion. Let us now look further afield.

Fig. 11.10. Comet NEAT. (Photograph by Martin Mobberley on January 2003).

Chapter 12

The Stellar Sky

When men of ancient times looked up into a starlit sky, they could see many hundreds of tiny, twinkling points that seemed to be arranged in definite patterns. It was natural, then, for the stars to be grouped into definite "constellations", each named after a deity, a demigod, or else some common object. Orion the Hunter, Hercules of legendary strength, and Perseus with the Gorgon's Head mingle with the Dragon, the Fishes, and the Cup. Forty-eight separate constellations are listed in the great catalogue contained in Ptolemy's *Almagest*, and may therefore be said to date from the dawn of astronomy.

The names are generally used in their Latin forms, so that the Dragon is "Draco" and the Fishes "Pisces". Any amateur who means to do serious work in the field of stellar research should become accustomed to the Latin names, which are in any case easy to remember. A full list, with the English equivalents, is given in Appendix XV.

Ptolemy's 48 constellations are still used today, but others have been added since. Some of these new groups lie near the south celestial pole, so that they never rise in the latitude of Alexandria, and Ptolemy naturally knew nothing about them; others have been formed by taking pieces away from the original 48. Further proposed additions with barbarous names such as Sceptrum Brandenburgicum, Officina Typographica and Lochium Funis have been mercifully forgotten, though one of the rejected groups, Quadrans Muralis (the Mural Quadrant) has left a legacy in the form of the name of the annual Quadrantid meteor shower seen each year from January 3 to 5.

Probably the most famous of the constellations are Ursa Major (the Great Bear), Orion, and Crux Australis (the Southern Cross). Of these, Ursa Major lies in the far north of the sky, so that in England it never sets, while Orion is crossed by the celestial equator and Crux is so far south that it never rises in our latitudes. Stars which never set are termed "circumpolar", so that Ursa Major is circumpolar in England.

To give a full explanation of the apparent movement of the star-sphere would be rather beyond our present scope, but something must be said about the essential terms Right Ascension and Declination. Broadly speaking, these are the celestial equivalents of longitude and latitude on the Earth's surface, though there are certain important differences in detail.

Declination is reckoned in degrees north or south of the celestial equator, while the equator itself is merely the projection of the Earth's equator in the sky. Clearly the north celestial pole will have declination 90 degrees north (+90°), and Polaris, the Pole Star, with its declination of greater than +89°, is so close to the polar point that it always indicates the approximate north celestial pole. Observers in the southern hemisphere are not so lucky, since there is no bright star placed conveniently at the south polar point.

To anyone observing from the north pole of the Earth's surface, Polaris would appear to remain virtually overhead; its altitude above the horizon would be greater than 89°. At Greenwich (latitude N. 51½°) the altitude of Polaris is 51½°; on the equator (latitude 0°) Polaris has of course no altitude at all – in other words, it lies right on the horizon. South of the terrestrial equator, Polaris never rises, so that it will never be seen.

If you want to work out whether some particular star is circumpolar, or whether it is permanently invisible, all you have to do is to subtract your latitude from 90 degrees and then compare it with declination of the star. For example, consider my own observatory in Selsey, Sussex, where the latitude is 51°North. 90 – 51 = 39. Therefore, any star north of declination +39° will never set, and any star south of declination – 39° will never rise. I can never see the bright southern star Canopus, whose declination is – 53°, but I can always see Alkaid in Ursa Major, where the declination is 49°. From Invercargill in New Zealand (latitude 46°S) Canopus is always on view, Alkaid never. In Appendix 23 I have given the declinations of some of the brightest stars.

The point at which the Sun crosses the celestial equator, moving from south to north, is known as the Vernal Equinox, of First Point of Aries. The Sun reaches it around March 21 each year, when its declination is, naturally zero. Six months later it crosses the equator again, this time moving from north to south, at the Autumnal Equinox or First Point of Libra. The Vernal Equinox is to the sky what the Prime Meridian is to the Earth – but it is a point which has been chosen for us by Nature, whereas Greenwich was elected as the zero for longitude merely because the famous Observatory happened to have been built there.*

The angular distance of a star eastward of the Vernal Equinox is known as the star's right ascension. It can be given in degrees, but it is more usually measured in hours, minutes and seconds of time because such a method is more convenient.

To explain this, we must refer to the 'meridian' of any observing point on the Earth; it is the great circle on the celestial sphere which passes through the celestial pole and also the zenith, or overhead point of the place of observation. Clearly, a star on the meridian will be at its maximum height above the horizon, and is said to 'culminate'. The First Point of Aries must pass across the meridian at any place once in every 24 hours (sidereal time), and the difference between this time and the time of the star's meridian passage will give the right ascension of the star. For instance, the brilliant Sirius reaches the meridian 6 hours 45 minutes 08.9 seconds after the First Point has done so, indicating that the right ascension of Sirius is 6h 45m 08.9s.

---

*Greenwich was accepted as the zero meridian by international agreement almost 100 years ago. The only countries to object were (naturally) France and Ireland.

The slight shift of the celestial pole due to precession, described in Chapter 2, means that a star's right ascension and declination alter very gradually over the years. In this book the positions are given for the year 2000, and it will be a long time before the error becomes great enough to be worrying. Incidentally, the First Point of Aries has now moved into the neighbouring constellation of Pisces, the Fishes, though we still keep to the old name.

A telescope equipped with setting circles and clock drive can be swung to any desired right ascension and declination, also that as soon as the position of the target object is known it can be found at once. It is obvious that while the right ascensions and declinations of stars do not alter much except over long periods, those of the Sun, Moon and planets will change quickly.

Dividing the stars into constellations, naming the brightest of them, is enough for a rough classification. Most of the brilliant stars have proper names, such as Sirius, Canopus, Rigel, Vega and Capella. However, it would be a hopeless task to give special names to each star, and we have recourse to letters or numbers.

A system introduced by the German astronomer Johann Bayer has stood the test of time so well that it will certainly never be altered. On this system, each of the leading stars of a constellation are allotted a Greek letter, beginning with Alpha for the brightest object and ending with Omega for the faintest. In Aries the Ram, for instance, the brightest star is Alpha Arietis (Alpha of the Ram), the second brightest Beta Arietis, and the third brightest Gamma Arietis. Unfortunately the strict order is often not followed, so that the system has become rather chaotic. In Orion, Beta is the brightest star, followed by Alpha, Gamma Epsilon, Zeta, and then Kappa, with Delta an "also ran". A list of the Greek letters, with their English names, is given in Appendix 16.

This is all very well, but it can deal with the 24 principal stars in each constellation, which in some cases (such as Orion) is not nearly enough. Flamsteed, the first Astronomer Royal, preferred to give the stars numbers, beginning in each constellation with the star of least right ascension. Still fainter stars, not listed by Flamsteed, have been allotted numbers according to later catalogues, and the result is that each bright star has several designations; Rigel in Orion is known also as Beta Orionis and as 19 Orionis. As time goes on, the proper names of the stars are becoming less and less used, with the Greek letters and the numbers taking their places.

It is also necessary to have some scale of reckoning apparent brilliancy. This is done by classification into "magnitudes" but the scale sometimes causes confusion, since the lowest values indicate the most brilliant objects. Bright stars are of magnitude 1, and the faintest visible to the normal eyes without a telescope are of magnitude 6, while with powerful telescopes stars down to the 30th magnitude can be detected. Modern instruments known as "photometers" can measure the brightness of a star very exactly, and in catalogues the value is given to 1/100 of a magnitude. Polaris, for instance, is of magnitude 1.99, so that it may be regarded as a standard star of the second magnitude.*

A few stars are actually brighter than magnitude 1.0, so that they have values of less than unity; examples are Rigel (0.08) and Altair (0.77), Rigel being appreciably

---

*The magnitude of Polaris is very slightly variable.

the brighter of the two. Four stars – Sirius, Canopus, Alpha Centauri and Arcturus – have minus magnitudes. On the stellar scale, Venus at its brightness is of magnitude $-4\frac{1}{2}$, while the Sun is about –27. The magnitude scale is based upon a definite mathematical ratio, but this need not concern us at the moment.

The stars are of different luminosities, and are at different distances from us, so that our constellation groups are due to mere line of sight effects. In Ursa Major, for instance, one of the seven bright stars (Alkaid) is much more remote than the other six, while Polaris in Ursa Minor or the Little Bear, is twice as distant as Alkaid. Merely because two stars are in the same constellation, we need not suppose that they have any connection with each other. There is an easy way of showing this. If you look at a gatepost as seen against the background of a clump trees, you do not suppose that the gatepost has any real connection with the trees.

As described by our scale model on page 13, the stars are so remote that their distances are not easy to measure. The first reliable results were obtained by using the method of "parallax", which is interesting enough to be explained more fully, even though it is useless for any but the very nearest stars.

The best way to demonstrate parallax is to make a practical experiment with a pencil, holding it up in front of your face and looking at it with alternate eyes. First align the pencil with some object, such as a vase on the mantelpiece, using your right eye only. Now shut your right eye and open your left, keeping the pencil still. The pencil will no longer seem to be in line with the vase; it will seem to have shifted. If you know the distance between your eyes, and the angular amount by which the pencil appears to have shifted, you can work out the actual distance of the pencil by using fairly simple mathematics (Fig. 12.1). This apparent shift in position is a measure of the pencil's parallax.

Much the same principle can be used to measure the distance of a relatively near star seen against a background of more distant stars, but the base-line used has to be enormously long. Fortunately Nature gives us such a base-line; the Earth swings from one side of the Sun to the other in a period of six months, shifting 186 million miles in position (Fig. 12.2). If S is our "near" star, it will appear to be at position S1 in January, but at S2 in July, so that if we measure the angular shift we can find the distance. (The diagram given here is hopelessly over-simplified and out of scale.) The actual amount of the shift is so minute that it is hard to measure, while there are numerous corrections to be made. However, there is nothing complex in

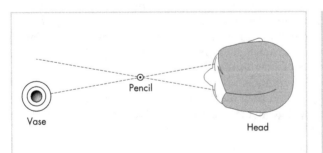

Vase

Pencil

Head

Fig. 12.1. Diagram to illustrate the principle of parallax.

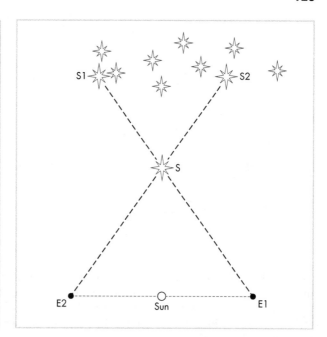

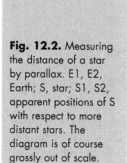

Fig. 12.2. Measuring the distance of a star by parallax. E1, E2, Earth; S, star; S1, S2, apparent positions of S with respect to more distant stars. The diagram is of course grossly out of scale.

the basic principle of the method, and it was in this way that Bessel managed to measure the distance of the fifth-magnitude star 61 Cygni, in 1838.

The parallax method breaks down altogether for all but the closer stars, because the shifts become too small to be properly measured. At 300 light-years the method has become untrustworthy, and at 600 light-years it is quite useless. Indirect methods have had to be developed, and most of these involve finding out the actual luminosity of a star as compared with the Sun, since as soon as we know the real brilliance and the apparent brilliance we can find the distance – much as we can judge the distance of a lighthouse if we know the power of the lamp and can measure how bright it appears to us.

Even the nearest of all the stars, Proxima Centauri in the southern sky, is immensely remote, so that in comparison even Pluto is very close at hand. The distance in miles is about 25 million million, or $4\frac{1}{3}$ light-years.

Sirius, which appears the brightest star in the sky, is 26 times as luminous as the Sun, but it owes its supreme position in our skies mainly to the fact that it is relatively close to us, since it lies at a distance of only $8\frac{1}{2}$ light-years. Canopus, in the southern constellation of Argo Navis (the Ship Argo), looks only a little less bright than Sirius, but is a great deal further away, so that it is clearly much more luminous. It would in fact take thousands of Suns to equal Canopus.

However, we must not imagine that our Sun is unusually feeble. It may be a firefly compared with Canopus, but it is a searchlight compared with some of the dimmest members of the stellar system. The faint red star known as Wolf 359 has a luminosity of only 1/50,000 of that of the Sun, so that we need not be too humble. If anything, the Sun is rather above the average in brilliancy, though there is really no such thing as an "average star".

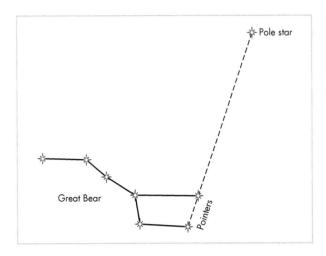

Fig. 12.3. The Pointers and the Pole Star.

Just as the stars are of different distances and luminosities, so they are of different sizes, temperatures and colours. A glance at Orion will show that of the two apparently brightest stars, one (Betelgeux*) is orange-red, while the other (Rigel) is white or slightly bluish. Betelgeux is the larger of the two, but it is much the less luminous, and its surface is cooler than that of Rigel. In fact, the stars present an almost infinite variety, so that it is rare indeed to find two which seem to be exactly alike.

To the ordinary observer, the stars appear to remain in fixed positions. Two of the stars in the Great Bear, Dubhe and Merak, always point to the Pole Star (Fig. 12.3); they have done so for generations, and will continue to do so for generations more. Of course, the old term "fixed stars" is misleading. The stars are moving about at high speeds, but they are so remote that it takes centuries for bright naked-eye stars to show obvious shifts in position, while the tiny annual shifts due to parallax can be detected only with the most refined instruments. Over the ages, however, the shifts will mount up, and eventually the two Pointers will no longer seem to line up with Polaris.

The slow movement of a star across the background is known as the star's Proper Motion, and must not be confused with the minute movement due to parallax. There is also a motion in the line of sight, termed Radial Motion (Fig. 12.4). If a star is coming straight towards us or away from us, it will have no proper motion at all, and will appear to remain still even over the lapse of centuries, but its radial motion will be detectable by means of the spectroscope.

Since the Sun is an ordinary star, the other stars show spectra of much the same kind. Temperature differences and other factors will cause complications, but usually there will be the continuous rainbow crossed by dark absorption lines due to gases in the star's reversing layer (page 64). If the star is approaching us, the dark lines will be shifted slightly towards the violet or short-wave end of the spectrum,

---

*This name may be spelled in various ways, such as Betelgeuse and Betelgeuze. In using a final x, I have followed the advice of Arabic scholars.

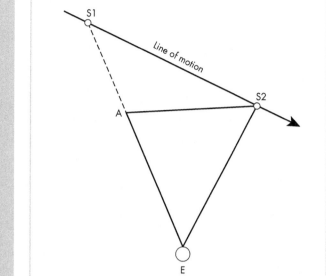

Fig. 12.4. Radial motion of a star. S1S2 = actual motion, S1A = radial motion, AS2 = proper motion.

while if the star is receding the shift will be towards the red. By measuring the amount of the shift, we can work out the radial velocity of the star.

There is an everyday analogy to this. When a train whistles, the whistle is high-pitched so long as the train is coming towards us, because more sound-waves are entering our ears, per second, than would be the case were the train standing still. After the train has passed by, and begins to draw away, fewer sound-waves will reach us per second, so that the pitch of the whistle drops. Light can be regarded as a wave-motion, and when the source of light is moving away the "pitch" is shifted towards the long-wave or red end of the spectrum. This is known as the Doppler Effect, in honour of the Austrian physicist Christian Doppler, who discovered it over a hundred years ago.

Sweeping the skies with a telescope is a fascinating occupation. Some of the stars show vivid colours; some are double, and some can be split into three or more components, so close together that to the unaided eye they appear as one star. There seems to be no end to it all, and no observer can hope to examine all the stellar wonders in the course of a lifetime. The more he sees, the more he must realize that our own Solar System is a minute speck in space.

Chapter 13

The Nature of a Star

Nearly everyone who uses a telescope for the first time expects to see a bright star such as Sirius or Rigel enlarged to a massive globe half filling the field of view. Disappointingly, this is not so. If the star looks like a definite disk, there is something wrong with the telescope – as not even the world's most powerful instrument, the VLT or Very Large Telescope in Chile, can show a star as a measurable disk. Of course the stars do look like disks when photographed, but this effect is purely photographic.

This is not because the stars are small. Some of them are big enough to hold the whole orbit of the Earth round the Sun. The small apparent size is due to the fact that they are so far away. Using indirect methods, the apparent diameter of the red star Betelgeux, in Orion, has been found to be 0.04 of a second of arc, and this is not very much. As most people know, a circle is divided into 360 degrees, each degree into 60 arc-minutes, and each minute into 60 arc-seconds; Betelgeux, therefore, shows 1/25 of an arc-second – yet its real diameter is over 200 million miles.

Most of our knowledge about the stars comes from the use of the spectroscope combined with an adequate telescope. Most stars show spectra basically similar to that of the Sun but there are wide variations in detail, and there are well-defined spectral types. These are summed up in Appendix 17 (page 208). The classes are given letters of the alphabet: W,O,B,A,F,G,K,M,R,N,S,L,T. Of these, L and T have been added recently to cater for very cool stars. For the rest, some thoughtful astronomer invented the mnemonic "Wow! Oh Be a Fine Girl – Kiss Me Right Now Sweetie", which is still used despite the howls of anguish from the Politically Correct crackpots. (Have you any ideas to accommodate L and T?)

The series is alphabetically chaotic because in some of the early schemes, various classes were found to be unnecessary (E and H, for example). Most stars belong to types O and M, but the series denotes decreasing temperature, W and O being the hottest and R, N and S the coolest (ignoring the recently-added L and T), The Sun belongs to Type G, near the middle of the list. A refinement is to divide each type into sub-grades from nought to nine, so A5 is midway between A0 and F0.

Some W and O stars, known as Wolf-Rayet stars in honour of the two astronomers who first described them in detail, have surface temperatures of over 35,000 degrees Centigrade, so that they are the hottest of the normal stars. Their spectra are peculiar, having in some cases a large proportion of bright lines instead of the usual dark ones, and they have set astronomers many problems, some of which remain to be solved. Most Wolf-Rayet stars are very remote, so that they appear faint in spite of their great luminosity, though two of them (Zeta Puppis and Gamma Velorum) are of the second magnitude; Gamma is too far south to rise in England.

Rigel in Orion has a B-class spectrum, and in fact all the leading stars in Orion are of this type, with the obvious exception of Betelgeux. The surface temperatures are in the region of 25,000 degrees Centigrade, so that B-stars are highly luminous. Somewhat less hot are the A-stars such as Sirius, with temperatures of about 11,000 degrees Centigrade; stars of type F, such as Canopus, are cooler still. Procyon is also of type F. Hydrogen and helium lines are less conspicuous, but calcium vapour is much in evidence.

The Sun is a typical G-type star, with a surface temperature of 6,000 degrees Centigrade. Here, of course, our investigations are helped by the fact that the solar spectrum can be studied in great detail. Another good example of a G-star is Capella, which appears as one of the most conspicuous stars in our skies.

The remaining types are orange (K) or orange-red (M, R, N and S), with temperatures ranging from 4,200 degrees down to only 2,000 degrees. Types N, R and S are comparatively rare and most of them are variable in brightness, while their spectra are complex and not at all easy to interpret. Acturus in Boötes is of Type K, while Betelgeux, Mira in Cetus and Antares in Scorpius belong to Type M.

It may be convenient to group the stars in this way, but we have only touched the fringe of the problem, Consider, for instance, two M-type stars, the brilliant Betelgeux and the dim Wolf 359. Betelgeux shines as brightly as 15,000 Suns, while Wolf 359 is a feeble body with only 1/50,000 of the Sun's candle-power, so have we any reason to class them together in the spectrum sequence? To say the least of it, they are ill-assorted companions.

One of the great discoveries of the early twentieth century was that apart from types W, O, B and A, the spectral classes tend to be separated into "giants" and "dwarfs". We can find many M-giants like Betelgeux, and many M-dwarfs like Wolf 359, but M-stars of intermediate luminosity are virtually absent. When it became possible to estimate the diameters of the stars, the distinction between giants and dwarfs became even more evident. Betelgeux is a vast globe about 200 million miles across, whereas Wolf 359 has a diameter of less than a million miles (Fig. 13.1). If we picture a scale model and make Betelgeux a globe with a diameter equal to that of a cricket pitch, Wolf 359 will be represented by a croquet ball.

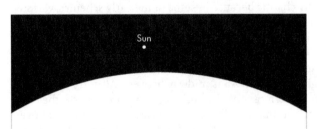

Sun

Fig. 13.1. Section of Betelgeux in Orion, showing size compared with that of the Sun.

The discovery of the giant and dwarf divisions was followed by a very simple, straightforward theory about the life-history of a star. It was assumed that in its early life, soon after it condensed out of the interstellar dust and gas, as star was hardly hot enough to emit visible light. Naturally, it would tend to shrink, because the force of gravity would tend to pull all its matter together; this would cause heat, so that the star would become a large Red Giant like Betelgeux. As the shrinking went on, the star would become an Orange Giant (type K) and then a Yellow Giant (type F), before turning into a smaller but very hot Wolf-Rayet or B star. As would be expected on this theory, the most luminous white types are not divided into giants and dwarfs.

This would be the peak of a star's career. It would go on shrinking, but it would also become cooler, since its main energy would have been spent. It would pass down the dwarf series or "Main Sequence", becoming first an F-dwarf, then a G-dwarf like the Sun, and then a small red star of one of the later types, finally losing all its heat and changing from a dim red dwarf like Wolf 359 into a cold, dead globe.

It all sounded beautifully straightforward – but, alas, it turned out to be completely wrong. The true sequence of events is much more complicated than this, and Red Giants such as Betelgeux are not young; indeed, they qualify as stellar OAPs.

Every star begins its career by condensing out of gas and dust inside a nebula. Inside nebulae we can observe small, dark patches known as Bok globules in honour of the Dutch astronomer Bart J. Bok, the first to draw attention to them. The denser regions inside them contract gravitationally and heat up, so that eventually they form masses of material known as protostars, and as time goes by these protostars draw in more and more material. What happens next depends entirely upon the initial mass of the protostar.

If the mass is less than 0.08 that of the Sun (80 times the mass of Jupiter) the star will simply glow feebly for an immense period before cooling down and ceasing to radiate; a star of this kind is called a brown dwarf. If the mass is higher, a true star will develop; the Sun belongs to this class. When the central temperature rises to 10,000,000 °C, nuclear reactions begin – and the main "fuel" is hydrogen, the most plentiful element in the entire universe (hydrogen atoms easily outnumber atoms of all the other elements obtained), and a normal star will contain a great deal of it. The nuclei of hydrogen atoms combine to make up nuclei of the second most plentiful element, helium. It takes four hydrogen nuclei to form one nucleus of helium; each time this happens a little energy is set free, and a little mass is lost. It is the energy which makes the star shine, and the mass-loss is considerable. For example, our relatively mild Sun loses "weight" at the rate of four million tons per second, so that it weighs much less now than it did when you began reading this page. However, please do not be alarmed. The Sun was born around six thousand million years ago, and as yet it is no more than middle-aged.

The Sun is now going through the most stable period of its career. Plot the stars on a diagram according to their spectra and luminosities, and you will find that most of them lie on a band known as the Main Sequence, extending from the top left of the diagram (hot stars of spectral types W, O and B-) down to the bottom right (cool stars of type M and later). A diagram of this sort is known as a Hertzsprung-Russell or HR Diagram, after the two astronomers who developed it – Ejnar Hertzsprung of Denmark and Henry Norris Russell of America.

Obviously the supply of available hydrogen "fuel" cannot last indefinitely, and when it begins to run low the star has to rearrange itself. The interior shrinks, while the outer layers expand and cool; the star leaves the Main Sequence and becomes a Red Giant. Different nuclear reactions begin at its core, and heavier and heavier elements are built up, but the star is living on borrowed time. For a while its luminosity increases at least a hundredfold (which means that when the Sun becomes a Red Giant, our luckless Earth cannot survive). The outer layers are puffed off to form what we call a planetary nebula; these disperse, and what is left of the star collapses to form a small, incredibly dense star of the type known as white dwarf. Sirius, the brightest star in our sky has a small white dwarf companion.

White dwarfs may be smaller than the Earth or even the Moon, but their masses are similar to that of the Sun, and it is only their small size which makes them look so faint. In a white dwarf the atoms are broken up, and their constituent parts are packed together with little waste of space. This explains the amazing density; a teaspoonful of white dwarf material would weigh several tons. The star cannot contract further, and will simply cool down until all its light and heat leaves it; it will soon become a dead star – a black dwarf. Yet the process takes a long time, and it is by no means certain that the universe is yet old enough for any black dwarfs to have been formed.

A star of greater mass will evolve more quickly, and die more violently. When the "fuel" runs out and collapse starts, the collapse is so catastrophic that nothing can stop it. Disaster follows. There is an "implosion", followed by an explosion; the constituents of the atoms fuse together to form particles called neutrons, and in a matter of seconds the star blows itself to pieces in a supernova outburst. The peak luminosity may be a thousand million times that of the Sun, and the density is unbelievable; you could pack a thousand million tons of neutron star material into an eggcup. Some rapidly-spinning neutron stars send out beams of radio radiation, and are called pulsars.

If the star is more massive still, it cannot even explode as a supernova. As the collapse proceeds the star pulls more and more strongly, until, not even light can escape from it; the star surrounds itself with a "forbidden area" from which nothing can break free – it has formed a black hole. Obviously we cannot see black holes, because they emit no radiation at all, but we can locate them because of their gravitational effects upon objects which we can see. We can do little more than speculate about conditions inside a black hole; all the ordinary laws of science break down.

Look out into space, and you are also looking back at time. For example, the Pole Star is 500 light-years away, so that its light, travelling at 186,000 miles per second, takes 500 yards to reach us; we see the Pole Star not as it is now, but as it was 500 years ago. Anyone living on a planet moving round the Pole Star would see the Earth as it used to be in Tudor times with Henry VIII as King of England – provided that our hypothetical observer has a sufficiently powerful telescope.

We have no idea whether there are planets attending the Pole Star, but we have found that planets of other stars do exist. They have yet to be seen, because they are too faint, but they can be detected in various ways, mainly by their gravitational effects on their parent stars. There is no reason to doubt that they support life, but as yet there is no way we can contact other civilizations – except possibly by radio. SETI, the Search for Extra-Terrestrial Intelligence, is now taken very seriously indeed.

I will say little about radio astronomy in this book, because I am dealing only with observation, and I am in no way a radio astronomer, but there is certainly room for amateur research in this field. A radio telescope is really in the nature of a large aerial, and collects radio waves just as an optical telescope collects light waves; no actual picture is produced – one certainly cannot look through a radio telescope! – but the information gathered is of vital importance in modern astronomy. Many people are familiar with the great 250 foot 'dish' at Jodrell Bank in Cheshire, but not all radio telescopes are 'dishes', because different lines of research need different techniques and instruments. Radio sources are of various kinds. The Sun is a powerful emitter, but many sources are supernova remnants, such as the Crab Nebula in Taurus, known to be the remnant of a brilliant supernova seen in 1054. There are radio galaxies, millions of light-years away, and the strange, super-luminous quasars – like pulsars, unknown and unsuspected when the first edition of this book was published. I will have more to say about them in Chapter 17. Meanwhile, let us turn back to the stars themselves.

Chapter 14

Double Stars

Of all the constellations in the sky, probably the best known is the Great Bear. It is not so brilliant as Orion, nor so spectacular as the Southern Cross; but it can always be seen in England when the night sky is clear, and most people have developed an affection for it. Besides, it is useful because two of its seven chief stars point to the Pole.

Even a casual glance will show something interesting about the "second star in the tail", known as Mizar or, on Bayer's system, Zeta Ursae Majoris. It is of the second magnitude, but close beside it is a much fainter star, Alcor, so dim that it is not particularly easy to see if there is the slightest haze.

Double stars of this kind are extremely common in the heavens, though most of them are too close together to be separated without the help of a telescope. They are spectacular enough to be well worth looking at for pure enjoyment, particularly when the two components are of different colours, and they are also useful for testing the performance of a telescope. A list of suitable "test pairs" is given in Appendix 20.

There are two classes of double stars. Sometimes the two members of a pair are not physically connected, so that the effect is due merely to the fact that one star happens to lie almost behind the other. One way to explain this is to picture two motor-cyclists coming down a long stretch of darkened road, using their head-lamps and separated by perhaps half a mile. An observer watching them approach may well imagine that they are riding side by side, particularly if the nearer cyclist has the less powerful lamp. However, "optical" double stars of this type are not so common as might be imagined.

The physical connection between the components of some doubles was first realized a century and a half ago by Sir William Herschel. Actually, Herschel made the discovery more or less by chance. He was trying to measure the distances of some of the stars by the parallax method (see page 122), and he had made long series of observations of pairs which he thought might show an annual shift. He failed in his main object, because his instruments were not sufficiently accurate; but he found that many of the doubles formed physically connected systems, and were in orbital motion round each other. Nowadays these genuine pairs are known as "binary stars".

It is not correct to say that the less massive star of a binary system revolves round its senior companion. Though the two may be very unequal in size and

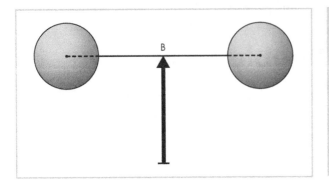

Fig. 14.1. The bells are equal in mass, and the balancing point B is midway between them.

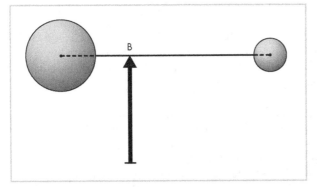

Fig. 14.2. The bells are unequal in mass, and B is no longer at the midpoint of the joining bar.

brilliance, they will certainly not be violently unequal in mass, since – as we have seen – the stars are strangely uniform in this respect; indeed, the smaller component may well be the more massive of the two. What will happen is that the two bodies will move round their common centre of gravity, much as the two bells of a dumb-bell move when twisted by their jointing arm.

If the two components have equal mass, the centre of gravity will lie half-way between them, just as we can balance the dumb-bells by the middle of the arm (Fig, 14.1). If one star is the more massive, the centre of gravity will be displaced towards it (Fig. 14.2). The Earth and the Moon move in this way; but since the Moon is so much the less massive, the centre of gravity of the system lies some way inside the Earth's globe.

A pair of binoculars will show that many apparently single stars consist of two, and a small telescope will reveal hundreds of pairs. Sometimes the components are equal, so that they are genuine twins, but more often one star is much brighter than the other. If a brilliant body is concerned, it may tend to drown its companion in a blaze of light, so that a telescope of some size will be needed to show both objects. Sirius is an excellent example of this. The brighter component is the most brilliant star in our skies, and it overpowers the White Dwarf companion, even though the White Dwarf would be an easy telescopic object were its shining on its own.

Binary stars have proved to be most useful to the theoretical astronomer. The orbits can be worked out; and as soon as the distance and the period of revolution are known, the combined mass of the stars in the system can be derived. Suppose, for instance, that the stars in a pair lie at an average distance from each other of 93 million miles, and have a period of one year. The Earth revolves round the Sun at this distance and in this time, and this means that the combined mass of the Sun-Earth pair must be equal to the combined mass of the two stars in the binary. In practice, we can neglect the Earth, which is of negligible mass when compared with any star, and in the above instance the two components of the binary would together equal one body the mass of the Sun. Unfortunately, calculating the separate masses of the components is not so straightforward.

The whole method depends upon careful measurements of the apparent relative motions of the twin stars, and it is therefore not surprisingly that most of the bright pairs have been so closely studied by professional astronomers that there is not much point in the amateur observing them further. Yet some of the generally-accepted measures are out of date, and there is definite scope for the serious observer with adequate equipment.

The separation of a double star is measured in seconds of arc. When it is borne in mind that the apparent diameter of the Moon is about half a degree, or 1,800 seconds of arc, it is evident that a pair of stars with a separation of only a second or two will need a powerful telescope if it is to be split. The apparent distance between Mizar and Alcor is roughly 700 seconds, but when a telescope is used the bright star is itself seen to be double, made up of two components between 14 and 15 seconds of arc apart. Actually, the system is more complicated than this.

There is a minor mystery connected with Alcor. The old Arab astronomers called it "a test for keen eyes", but nowadays it can be seen by any normal person when the sky is clear, and it can in no sense be regarded as a test. Either Alcor has brightened up during the last thousand years, or else it is not the star referred to by the Arabs. The real test star may be the much fainter object lying between Mizar and Alcor. This star is usually below the 8th magnitude, and thus quite invisible without a telescope, but it has been suspected of variability.

The "position angle" of a double star, binary or otherwise, is the direction of the fainter star as reckoned from the brighter, beginning with 0 degrees at the north point and reckoning round by east (90 degrees), south (180), and west (270) back to 0, as shown in Fig. 14.3. This is generally enough to enable one to form a mental picture of the double before one actually goes to a telescope, though in case of perfect twins it is not easy to tell which of the components is meant to be the senior partner.

Measuring the separations and position angles of double stars cannot be undertaken with a telescope of less than 6 inches in aperture, and it is also necessary to have an equatorial mount, a driving clock, and a measuring device known as a micrometer. Micrometers are of various types; to describe them here would be beyond our scope, but full information can be found in the works listed in Appendix 30.

The most beautiful of the double stars are those which show contrasting colours. Pride of place must go to Albireo or Beta Cygni, the faintest star in the "cross" of Cygnus, the Swan (Map VIII). The main star is of the third magnitude, and is of a strong golden-yellow colour, while the fifth-magnitude companion is a glorious blue-green. The two are sufficiently wide apart to be well seen in a 2-inch telescope, and a power of 50 on a 3-inch refractor will show them excellently. Other yellow and green pairs are known, but in my opinion, at least, none can rival Albireo.

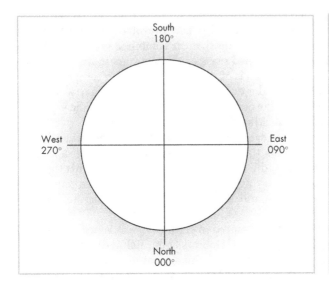

Fig. 14.3. Position angle.

There are also cases of bright orange-red stars, usually of Type M, which are accompanied by small green companions. Antares, leader of the Zodiacal constellation of Scorpius, is one of the reddest of the brilliant stars – its very name means "Rival of Mars" – and its beauty is enhanced by the fact that a small telescope will reveal an emerald-green star close beside it. The greenness of the faint companion is due partly to contrast with the ruddy hue of the giant, but it is nonetheless spectacular for that.

Now and then we meet with some oddly-assorted pairs. One of the most interesting is Sirius, the Dog-Star (Map V). The main component is an A-type giant with a luminosity 26 times as great as the Sun's, and a diameter of more than a million miles. The second star could hardly be more dissimilar; it is a White Dwarf, considerably

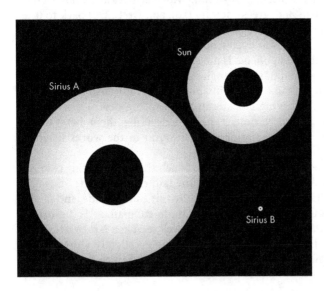

Fig. 14.4. Comparative sizes of Sirius A, Sirius B and the Sun.

smaller than Uranus, but with a mass almost equal to that of the Sun. In Fig. 14.4, the sizes of the two companions are shown, with the Sun added for comparison.

Since Sirius has always been known as the Dog-Star, the White Dwarf companion has acquired the nickname of "the Pup", but at least it is a pup which can make its presence felt. Like Neptune, it was tracked down by its gravitational pull long before it was actually seen. Bessel, famous as the first astronomer to measure the distance of a star, found that Sirius itself was wobbling slightly in the heavens, and he calculated that this must be due to the effect of an unseen companion. Years later, in 1862, the Pup was discovered, quite by chance, by an American instrument-maker who was testing a large new telescope.

Though the two stars of the Sirius pair are so unequal in size and luminosity, the bright giant has a mass only 2½ times as great as that of the White Dwarf. The distance between the two is about equal to that between Uranus and the Sun, and the period is about fifty years, so that two complete revolutions have been completed since the Pup was first seen. As a matter of fact, the Pup does not appear to be particularly faint, but it is not easy to observe, since the glare from the larger star drowns it. It has been claimed that a 6-inch telescope will show it, but I admit that I have yet to see it with my 12½ inch reflector, probably because Sirius lies well south of the celestial equator and so never rises high above the horizon in England. However, I have seen it with my 15-inch reflector, and keener-eyed observers have reported it with much smaller telescopes.

Some double stars are too close to be split with any telescope, but can nevertheless be detected by means of our old and reliable ally, the Doppler Effect. In the very much over-simplified diagram given in Fig. 14.5, it is assumed that the fainter star (B) is revolving round the brighter (A). In position 1, B is moving towards us, and its spectrum will show a violet shift; in position 2, it will be receding, and the shift will be towards the red. Consequently, the combined spectrum due to the two stars will show variations, and the binary nature of the system will be betrayed. Even if the spectrum of one component is too faint to be seen at all, the wobbling of the lines of the other star will be just as tell-tale. Pairs of this kind are termed "spectroscopic binaries".

Fig. 14.5. The Doppler Effect for spectroscopic binaries.

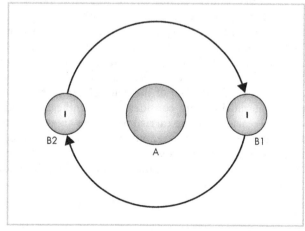

Fig. 14.6. The famous "double-double" star Epsilon Lyrae.

Now and then we meet with positive family parties of star systems including three, four or even six components. One of the best known is Epsilon Lyrae, shown in Map VIII, lying close to the brilliant star Vega, which appears almost directly overhead in England during summer evenings. Keen eyes can see that Epsilon is made up of two components, and in binoculars the pair can be well seen, since the apparent distance between them is 207 seconds of arc. A 3-inch telescope reveals that each component is again double, so that there are four visible stars in the system (Fig. 14.6), and to make things even more complex one of the four is itself a spectroscopic binary. The two main pairs are so far apart that they take at least a million years to complete one revolution around their centre of gravity.

Equally remarkable is Castor, one of the main stars in the famous constellation of Gemini, the Twins (Map V). Here we have two bright components at present 1.8 seconds of arc apart, though the revolution period is 380 years and it is not now so easy to split the pair as it used to be half a century ago. Each is a spectroscopic binary, and there is a 9th-magnitude spectroscopic binary companion 73 seconds of arc away, so that the system of Castor is made up of six separate suns. On the other hand, Gamma Virginis, in the Y of Virgo (Map VI) used to be a grand, easy pair with equal components, but we now see it from a less favourable angle, and it has become difficult, though in the future it will 'open out' again.

The magnification for looking at any particular double star must depend upon the individual double itself. If you want to obtain an overall view of Mizar and its companions, a low power is necessary, since if you increase the magnification you will find that Alcor is out of the field. Closer pairs naturally need higher powers, and for measuring work considerable magnification must be used.

Useful research can be carried out by the amateur double-star observer; there is still routine work to be done, and in any case there is much enjoyment to be gained from looking at the pairs and groups of suns. With their varied separations and their lovely contrasting colours, they are among the most beautiful of the objects in the stellar heavens.

Chapter 15

Variable Stars

Fortunately for us, our Sun is a steady, well-behaved star. It may have periods of unusual activity, when its disk is disturbed by spot-groups and flares, but at least its output of energy does not alter greatly over the lapse of hundreds of centuries.

Other suns are not so quiescent. Some of them vary in brightness from day to day, even from hour to hour, either regularly or in an erratic manner. They swell and shrink, and their temperatures change with their fluctuations, so that any planet circling round them would be subject to most uncomfortable changes of climate.

Variable stars are important both to the professional and to the amateur, and the owner of a small instrument can do useful work, particularly as his telescope need not be so perfect as that of the lunar or planetary observer (though, of course, the better the telescope the better the results). It is true that the regular variables of short period have been closely studied at the great observatories, but there are other stars which seem to delight in springing surprises, so that they need constant watching.

Variable stars are of many types, but it is not difficult to give a general classification.

First there are the eclipsing binaries, such as Algol in Perseus, which are not true "variables" at all, even though they do seem to alter in brightness. Perhaps the most important of the true short-period variables are the Cepheids, so named because the star Delta Cephei is the best-known member of the class; the periods range from a few days up to six or seven weeks. Of much shorter period are the RR Lyrae stars, whose periods range between 30 hours and less than 2 hours. Then there are the long-period variables, usually Red Giants of great size and comparatively low temperature, with periods ranging from 70 days to over 2 years. Irregular variables, as their name suggests, behave in an unpredictable manner. Lastly come the violently explosive "temporary stars" or novae.

There are several variables which can be followed without any telescope at all. The most famous of these is Betelgeux, the Red Giant in Orion. It belongs to the irregular class, though there is a very rough period of from 4 to 5 years, and it changes in brightness from magnitude 0 down to 1, so that whereas it may sometimes almost equal the glittering Rigel it may at others be comparable with Aldebaran, the "Eye of the Bull;". The alterations are slow, but they become noticeable over a week or two, and the beginner who estimates the magnitude of

Betelgeux every few days will soon be able to detect the fluctuations. However, most of the interesting variables cannot be followed without a telescope or at least binoculars, since when near minimum they are below naked-eye visibility.

Before coming to the proper variables, it will be of interest to say something about the "fake variables", or eclipsing binaries. These might well have been described in the chapter dealing with double stars, but since they do seem to change in brilliancy they come under the scope of the variable star enthusiast.

The best-known of these "fakes" is Algol, which lies in the constellation of Perseus and is shown in Map VII. In mythology, Perseus was the hero who slew the fearful Gorgon, Medusa, whose glance turned the hardiest onlooker to stone, and it is fitting that Algol should mark the Gorgon's severed head.

Usually Algol shines as a star of magnitude 2.1, only a little inferior to Polaris. It remains constant (or virtually so) for a period of $2\frac{1}{2}$ days, but then it starts to fade, until after about five hours it has dropped to magnitude 3.3. After a relatively brief minimum, it starts to brighten once more, taking a further five hours to regain its lost lustre. Textbooks usually say that its variations were discovered by Montanari in 1667, but the old Arab astronomers called Algol "The Winking Demon", which is interesting if they were unaware of its odd behaviour – as they seem to have been.

Algol is not truly variable. The apparent fluctuations are due to the fact that the system is a binary, and when the brighter star is eclipsed by the fainter the total brightness naturally drops. When the fainter star is obscured by the brighter, there is a small minimum, but since this amounts to only one-twentieth of a magnitude it cannot be detected with the naked eye. Actually the system of Algol includes a third star, but the principle of the variations is straightforward enough.

The beginner may like to plot Algol's "light curve". A light-curve is merely a graph plotting time against magnitude, as shown in Fig. 15.1, and it is always interesting to make one from personal observations. In the diagram of Algol given here, the secondary minimum is slightly exaggerated, as otherwise it would not be visible upon a chart drawn to so small a scale.

Another bright eclipsing binary is Beta Lyrae, which lies near the brilliant Vega (Map VIII). Here are two bright components, so close together that they almost touch, and in consequence too close to be seen separately in any telescope. At maximum, when both stars are shining together, Beta Lyrae appears of magnitude 3.4. It then fades steadily to magnitude 3.8, and then rises once more to 3.4, but at the next minimum it descends to 4.4, so that deep and shallow minima take place alternately. The brightness is always varying, so that there is no long comparatively steady maximum, as with Algol. One remarkable fact about the components of Beta Lyrae is that each is stretched out into the shape of an egg, simply because the two stars are pulling so strongly on each other; the general situation has been compared to two eggs rolling about with their sharper ends kept close together.

Epsilon Aurigae is also an eclipsing binary (Map IV), though its nature is still problematic. The period is over 27 years, the longest known for any eclipsing star. Its neighbour in the sky, Zeat Aurigae, has a period of 972 days, and is particularly interesting to spectroscopic workers because the smaller star shines for some time through the outer layers of the diffuse giant component before disappearing behind. The fluctuations of Epsilon and Zeta Aurigae are however much less obvious than those of Algol, and are not marked enough to be noticeable with the naked eye.

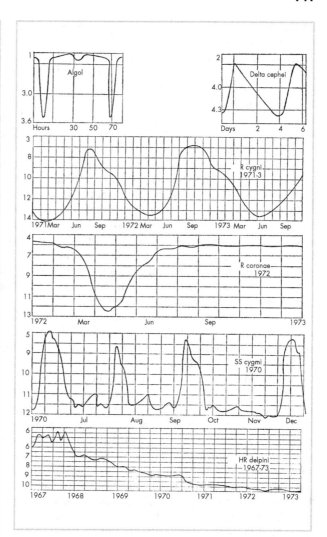

Fig. 15.1. Light-curves of variable stars. (1) Algol: eclipsing binary. (2) Delta Cephei: prototype Cepheid. (3) R Cygni: Mira type. From my observations of 1971–1973 (12½-inch reflector). (4) R Coronae, 1972: a typical minimum. Between August 1972 and October 1973, the magnitude remained almost constant at 6 (12½-inch reflector). (5) SS Cygni, 1970: Again I used my 12½-inch reflector. Four maxima occurred. (6) HR Delphini, 1967–1973: Binoculars till 1969, then my 12½-inch reflector. (The slowest nova on record!)

Turning now to genuine variables, we must begin with the Cepheids, which are of great importance because they are obliging enough to act as "standard candles". Several are visible without a telescope, the best known being of course Delta Cephei itself, which lies fairly close to the north celestial pole (Map VII) and therefore remains permanently above the horizon in England. The period is 5½ days, with a magnitude range of from 3.5 to 4.4, and the light-curve is not symmetrical; the rise from minimum to maximum is quicker than the subsequent fall, and this is always the case, since Delta Cephei's variations are so regular that the period is known to within a fraction of a second.

A Cepheid seems to be a pulsating star, expanding and contracting rather in the way that a balloon will do if air is forced in and out of it. This is no mere theory; it has been proved, not by the telescope but by the Doppler Effect. When a Cepheid is expanding, its bright surface is moving towards us, and the lines in the spectrum are shifted towards the violet; when the star is contracting, the surface is receding from us, and the shift is towards the red. In general, Cepheids have a small magnitude-range (the Pole Star is actually a Cepheid, though its fluctuations are too slight to be detectable with the naked eye) and spectroscopic studies of them are of great importance.

Equally important is the Period-Luminosity Law, which has provided the stellar astronomer with one of his most powerful weapons. Reduced to its simplest terms, this Law links the variation period of a Cepheid with the star's actual luminosity, so that variables of equal period have the same candle-power. Delta Cephei, period $5\frac{1}{2}$ days, is approximately 660 times as luminous as the Sun; therefore, every Cepheid with a period of $5\frac{1}{3}$ days is 660 times as luminous as the Sun.

This by itself would be intriguing enough, but it has far-reaching consequences. If we know the real brightness of a distant lighthouse, and we can measure how bright it appears to be, we can work out its distance from us by means of simple arithmetic. In the case of Delta Cephei, we know its real luminosity and its apparent magnitude, so that its distance follows at once; it proves to be 1300 light-years. In fact we can find out the distance of any Cepheid merely by watching how long it takes to vary from maximum to maximum.

The Law has been known now for seventy years and there is no doubt of its validity. The longer the period, the more luminous the star. These strange variables are the standard candles of the universe, and they never depart from their own rules, even though we know them to be less uniform in type than used to be believed.

Consequently, we find that we have the means of measuring the distance of a remote star-cluster or galaxy. If we can detect a Cepheid, we can find its distance, and so the distance of the cluster which contains it must be the same. Nature can be awkward at times, as we know to our cost, but in any case she has given us an expectedly accurate measuring-rod. There are certain complications, since there are two different types of Cepheids with different period-luminosity relationships, but refinements of this nature do not seriously mar our space-gauging.

RR Lyrae stars were formerly classed with the Cepheids, but it now appears that they form a separate group. Their fluctuations are perfectly regular, and their periods range from only $1\frac{1}{2}$ hours up to slightly more than a day. All RR Lyrae stars have about the same luminosity, roughly 95 times that of the Sun, so that they too can be used as standard candles. All are distant, and so appear too faint to be easily studied by the amateur observer.

The periods of the eclipsing binaries, the Cepheids and the RR Lyrae stars, are known so accurately that there is no point in a the amateur's observing them further. Nor do any reasonably bright variables of such types remain to be discovered. On the other hand, the long-period stars present very different problems. They are not perfectly regular, and they are not so closely studied by professional astronomers, so that here the amateur can come into his own.

In August 1596, David Fabricus recorded a third-magnitude star in the constellation of Cetus the Whale, not far from Orion (Map IV). By October it had disappeared. Bayer saw it again in 1603, when he was drawing up his star catalogue, and gave it the Greek letter Omicron, but shortly afterwards it vanished once more. Not until some time later was it found that the star appears with fair regularity; it takes approximately 331 days to pass from maximum to maximum, and it is visible to the naked eye for many weeks at a time. Not unnaturally, it was given the name of Mira, "The Wonderful".

The period of naked-eye visibility is not always the same, and nor is the magnitude at maximum. At some maxima, as in 1969, the star attains the 2nd magnitude, and remains visible without a telescope for over 20 weeks, but in other years it becomes no brighter than magnitude 5. In 1868, for instance, it was a naked-eye object for only 12 weeks. Near minimum the magnitude falls to below 9, so that Mira cannot then be found even with binoculars or a small telescope. Nor is the period constant; the 331 days given in most textbooks is merely an average, and may fluctuate to the extent of more than a month either way. There is nothing neat or precise about "the Wonderful Star", and for this reason alone it is worth keeping under watch. Extra interest is added by the tiny white companion, so faint that it is hard to see except when the senior star is near minimum.

Like all long-period variables, Mira is a Red Giant, large, cool and diffuse. Many similar stars are known, some of which can be seen with the naked eye when at their brightest, and our knowledge of their behaviour depends mainly upon the results of amateur work. There is no period-luminosity law, and thus the stars cannot be used as standard beacons, with the result that professional astronomers do not study them so closely as in the case of Cepheids. Again there is a pulsation as well as a change in temperature, but the whole behaviour of the stars is different. It is rather strange to find that stars with the longer periods often prove to be of relatively low luminosity.

Though the long-period stars are at least partly regular, there are some variables which seem to be completely erratic. These irregular variables are perhaps the most fascinating of all, since one never knows what they are going to do next. Betelgeux is one example, and other Red Giants which behave in a similar way are Alpha Herculis (Map IX) and Mu Cephei (Map VII). Mu Cephei is particularly interesting. It varies between magnitudes 3.6 and 5.1, so that it can always be seen with the naked eye, but a pair of binoculars will show that it is of a beautiful red colour. It looks almost like a drop of blood, and it deserves the name of "the Garnet Star" given to it by Sir William Herschel.

Cassiopeia, the Queen, is one of the most prominent of the northern constellations, and few people can mistake its five chief stars, which are arranged in the form of a rough W (Map VII). The middle star of the W, Gamma Cassiopeiae, is an interesting variable. It used to be ranked as a steady body of magnitude 2.3, but in 1936 it abruptly brightened up by over half a magnitude, so that it far outshone Polaris. Since then it has varied between magnitudes 2 and 3.3. Its spectrum is so peculiar that it cannot be placed in any other type.

Telescopic irregular variables are of many types. For instance, R Coronae, in the Northern Crown, is generally of about the 6th magnitude, on the fringe of naked-eye visibility, but at irregular intervals it drops down sharply, and fades to perhaps the 14th magnitude. Stars such as SS Cygni and U Geminorum remain at minimum for most of the time, but show sudden increases of several magnitudes; SS Cygni itself is usually of about magnitude 12, but can rise to above 9. R Scuti, in the little constellation of the Shield, has alternate deep and shallow minima, but sometimes loses all semblance of regularity for a while. And Eta Carinae, in the far south, is completely irregular; for a while, between 1837 and 1854, it ranked among the most brilliant stars in the sky, but for many years now it has remained below naked-eye visibility. On the whole, it is the irregular and semi-regular stars which offer the greatest scope for amateur observers, if only because one can never tell just what they will do next.

Variable star observations are made by estimating the magnitude of the variable as compared with near-by stars of known brightness. For instance, Gamma Cassiopeiae is provided with two perfect comparison stars in the same constellation, Beta (magnitude 2.26) and Delta (magnitude 2.67). In the case of a telescopic object, the comparison stars must of course lie in the same field as the variable, and a few awkward stars which lie aloof by themselves are not easy to estimate properly.

The first thing to do is to identify the variable. A star atlas is necessary, probably together with a chart type similar to those in Appendix 26, and the position of the variable can be found. It is, however, a mistake to look directly for the variable itself. The best method is to note the stars which will be found in the same low-power field, so that an overall impression can be built up. Most long-period variables stand out because of their redness, but this is never a safe guide, and is in any case not valid for the short-period stars and the irregulars.

It may sound difficult to identify any particular starfield, but no two fields are alike, and a little practice will work wonders. It is sometimes suggested that the best way is by "sweeping about" until the required field comes into view, but this is a mistake. When a telescopic variable is to be sought, there should be a definite plan of campaign. First identify the area by means of naked-eye stars which can be recognized without possibility of error, and then proceed by means of star alignments and patterns, swinging the telescope north and south in right ascension in terms of a known angular field. In difficult cases, an easily recognizable star can be selected which has the same declination as the variable, and the telescope left stationary until the variable drifts into view (though slight adjustments will be needed if the telescope is mounted on an altazimuth stand). It is unwise to leave any "safe anchorage" for the next until it has been identified with absolute certainty. When an observer has once found the field, he will usually recognize it again without much trouble, and it can be picked up in a matter of seconds, but the approach should always be "planned". A moment's carelessness can lead to some very peculiar results.

If the observer belongs to an astronomical society, he can of course obtain charts of the fields he needs. Approximate positions of some of the long-period and irregular variables are shown in the star maps given on pages 221–272.

There are several methods of making estimations. One of the simplest is Pogson's Step Method, in which the observer trains himself to gauge a difference of 0.1 magnitude, which constitutes one "step". Suppose that he is observing a variable star, and finds that it is two steps fainter than comparison star A and one step brighter than comparison star B. He records: "A-2: B+1". He then looks up the magnitudes listed for A and B. If A is 8.0 and B 8.3, the variable must be 8.2, which is two-tenths of a magnitude fainter than 8.0 and one-tenth of a magnitude brighter than 8.3.

A more complex method is the Fractional, used by many workers. Here two comparison stars are used, and the brightness difference between them is divided mentally into a convenient number of parts, after which the variable is placed in its correct position in the step-series. If A is the brighter of two comparison stars A and B, and the variable is estimated as one-quarter of the way from A to B (and hence three-quarters of the way from B to A), the record will read: A1V3B. The magnitudes of the comparison stars can then be looked up as before, and the magnitude of the variable worked out.

There are many points to bear in mind when using either of these methods, and perhaps the most important is that the observer should go to his telescope with an open mind. If he expects the variable to be of magnitude 7.5, there is a strong chance that he will in fact record it as 7.5, whether this is correct or not! Neither is it easy to compare a red star with a white one. Plenty of practice is needed, but the serious enthusiast will soon find that he has "got the hang of it", after which he will be able to estimate many variables during the course of a few hours' work.

One difficulty of observing naked-eye variables such as Betelgeux is that a star is bound to be reduced in brightness as it approaches the horizon, since it will be shining through a thicker layer of the Earth's atmosphere. This "extinction" effect can upset an observation completely if it is not allowed for, but the table given in Appendix 21 should help. With telescopic observations, extinction can be neglected, since all the stars in the field will be approximately the same altitude above the horizon.

Come now to the novae, which can always take us by surprise. Occasionally a star will blaze up where no bright star has been seen before, remaining prominent for a few days or weeks before fading back to obscurity. Since nobody knows when or where to expect them, amateurs have a splendid reputation as nova-hunters.

An ordinary or "classical" nova is a binary system, made up of a low-density red star together with a white dwarf companion. The white star pulls material away from its neighbour, and this material builds up round the white dwarf. Over a long period so much material accumulates that the situation becomes unstable, and a runaway nuclear reaction begins, with material hurled away violently into space. The outburst is brief, and the nova soon returns to its old state. The maximum luminosity during the outburst may be well over a million times that of the Sun, making ordinary variables seem very tame.

Over thirty naked-eye novae and many fainter ones have been seen since 1900. Pride of place must undoubtedly go to Nova Persei 1901 and Nova Aquilae 1918, each of which became brighter than any stars in the sky apart from Sirius and Canopus, but which have by now become very faint telescopic objects. Of particular interest was Nova Herculis 1934, which, like many other novae, was discovered by an amateur observer. It was found on December 13 by J. P. M. Prentice, then Director of the Meteor Section of the British Astronomical Association, who had been observing shooting-stars and was taking a nocturnal stroll after finishing his regular programme. The star had an unusually long maximum, and as it faded it developed a strong greenish hue, which was most striking with the 3-inch refractor that I was using at the time.

Novae generally appear near the Milky Way zone, but they are quite unpredictable, and the increase in light is usually so rapid that the amateur sky-watcher has a better chance of making the discovery than his professional colleague who is busy with a set programme of work. Novae can of course be estimated in the same way as normal variables, and it is fascinating to watch them as they fall gradually from their pinnacle of glory back to the obscurity from which they came.

Normal novae are spectacular enough, but the rarer "supernovae" are even more so. Here the increase in light is much greater, and at maximum a supernova may shine as brightly as all the other stars in its system put together. In our own galaxy, the most famous supernova on record is that of 1572. It lay in Cassiopeia, and at its brightest was more brilliant than Venus, so that it remained visible in broad daylight. Telescopes still lay in the future, so that as soon as the star fell below the sixth magnitude it could not be followed, but it has left a remnant which was first identified in modern times because it is a radio source. The Crab Nebula – the remnant of the supernova of 1054 – is a particularly powerful radio emitter, and contains a pulsar which shows up as a very faint object, flashing 30 times per second. The only other supernovae seen in our own system over the past thousand years have been the stars of 1006 and 1604. Supernovae are of two main types, one involving the complete destruction of the white dwarf component of a binary system, and the other due to the collapse and subsequent explosion of a very massive star. Supernovae are often seen in external galaxies, and in 1987 one appeared in the Large Cloud of Magellan, a galaxy only 169,000 light-years away. For a while it was almost as bright as the Pole Star, but it was too far south in the sky to be seen from anywhere in Europe.

Normal bright novae may appear at any moment, and the amateur will often make a naked-eye or binocular survey of the Milky Way zone. I remember the nova of 1967, HR Delphini, which was found late at night by George Alcock, an amateur astronomer who lived in Peterborough. He telephoned me, and asked me to confirm it; I did so at once – it was of the fourth magnitude, and remained near

Fig. 15.2. Supernova 1987a in the large cloud of Magellan. I took this picture with an ordinary camera. The supernova, then about as bright as the Pole Star, is in the centre of the picture.

Fig. 15.3. SN 2003 in NGC 4051. (Photograph by Martin Mobberley on 23 September 2003).

maximum for months. On the other hand the nova which flared up in Cygnus in August 1975 reached the second magnitude, but remained a naked-eye object for less than a week; it has now become very faint indeed, whereas HR Delphini has simply reverted to its pre-outburst magnitude of just below 12.

Obviously, visual estimates of variable stars can never be as accurate as measurements with photoelectric equipment, but they are still very useful, and are warmly welcomed by professional astronomers. There is plenty of scope.

Chapter 16

Star Clusters and Nebulae

Some way from Orion, beyond the bright red star Aldebaran, can be seen what at first sight looks like a faint misty patch. Close inspection shows that this patch is in fact made up of stars, one of which is of the third magnitude and the rest much dimmer.

Seven stars can be made out by normal-sighted people, and the group is known popularly as the "Seven Sisters",* though its official name is the Pleiades. It is a genuine cluster, and not a line-of-sight effect. It has been calculated that the odds against any chance of alignment of the seven most conspicuous stars are millions to one against.

The Pleiades have been known from very early times, and legends about them are found in ancient mythology, but it is only during the last three and a half centuries that astronomers have realized that there are many similar clusters in the sky. One or two can be seen without optical aid; there are the Hyades round Aldebaran, Praesepe or the "Beehive" in Cancer, and the Sword-Handle in Perseus. Most, however are too faint to be seen without a telescope.

A pair of binoculars will show the Pleiades very well. With a magnification of about 20, the chief stars fill the field, and look like jewels gleaming against black velvet. Moreover, fainter stars jump into visibility; even a small telescope reveals so many that to count them would be a very difficult process. The Seven Sisters have many junior relatives – over 250, in fact.

The Pleiad stars look close together, but the cluster is not really so dense as might be imagined, though if the Sun lay in the middle of the Pleiades our sky would contain many stars shining more brightly than Sirius does to us. Nor must we be deceived by the fact that the whole cluster takes up only a small patch of the heavens, since the real diameter of the group is over 15 light-years. Its distance is over 400 light-years.

Almost as famous as the Pleiades are the Hyades, which lie around Aldebaran itself, and are shown in Map IV. Actually, Aldebaran is not a genuine member of the cluster, as it merely happens to lie in the same direction. Telescopically the Hyades

*Really keen-sighted people can see up to a dozen Pleiads without optical aid, but artificial lights make it difficult to see the cluster as anything but a dim glimmer.

are not so beautiful as the Seven Sisters, as the stars are much wider apart, and it is difficult to get them all into the same field of view. Moreover, they are overpowered by the bright orange-red light of Aldebaran.

There is an important difference between the two clusters. In the Pleiades, the brightest stars are blue, highly luminous giants of type B, whereas in the Hyades the chief members of the group are orange giants of type K. B-stars do occur in the Hyades, but are much less in evidence.

Another naked-eye "open cluster" is Praesepe, shown in Map V. It lies in Cancer the Crab, and has been nicknamed the Beehive, because in a small telescope it has been said to give some impression of a collection of luminous bees. It is not prominent without a telescope, and even a half-moon is enough to drown it, but it is a fine sight in a small instrument (Fig. 16.1).

Fig. 16.1. The Pleiades. (Photographed by the author with an ordinary camera.)

Even more striking are the twin clusters in Perseus (Map VII), marking the "sword-handle" of the legendary hero. To the naked eye the only indication of their presence is an ill-defined misty patch, but a telescope reveals two rich star-clusters in the same low-power field. I have found that good view is obtained with a power of about 30 on a 3-inch refractor (Fig. 16.2).

Telescopic clusters are numerous, and anyone equipped with a small instrument can give himself many hours of enjoyment by sweeping for them and learning how to pick them up. Each has a separate designation, most of the brightest being known by their numbers in Messier's catalogue. Charles Messier, it will be remembered, was the French comet-hunter who was constantly annoyed by confusing clusters with comets, and so drew up a list of objects to be avoided during his searches. Thus Praesepe is M.44, the Pleiades M.45, and the nebula in Orion M.42. The full catalogue, given in Appendix 26, contains 107 objects. A few of the objects listed by Messier cannot now be found, and may have been comets that the French observer failed to recognize for what they were.

Because Messier was not actually interested in nebulae, he catalogued only those which could be confused with comets – and of course he had to leave out much of the southern sky, because he spent all his life in France. Some years ago I drew up a catalogue of over 100 objects not listed by Messier, and published it – as the

Fig. 16.2. M44 Praesepe. (Photograph by Steve Barnes.)

Caldwell Catalogue (obviously I could not use M numbers, but my surname is actually a hyphenated one – Caldwell-Moore – so I used C*).

Praesepe and the Pleiades are open clusters, but both the catalogues include objects which are quite different – for instance the globular clusters, which look like compact balls of stars, so closely crowded towards their centres that it is difficult to distinguish the individual points of light. A rich globular may contain a million separate stars, and the crowding is much greater than in the case of the open clusters.

All globulars are very remote, and even the nearest of them lies at a distance of thousands of light-years. They form a sort of "outer surround" to our stellar system, and since the Sun is not in the middle of the Galaxy we naturally have a better view of the globulars to one side of the sky. Most of them appear round the southern constellations of Scorpius and Sagittarius.

The best way to demonstrate this effect is to imagine that we are standing in a woodland glade on a foggy evening. If we stand away from the centre of the glade, we can see the bordering trees to one side of us, but the trees which mark the edge of the glade on the far side will be concealed by the fog. If we take each tree to represent a globular, we can understand why these strange clusters are best seen in one particular direction. Space, too, is "foggy"; there is a good deal of inter-stellar dust and gas, and light-waves cannot penetrate it, so that in certain directions our view is blocked.

The globulars are too far away to have their distances measured directly, but fortunately they contain RR Lyrae stars, and these useful beacons give us the answers at once. As has been shown, we can find the distance of an RR Lyrae star simply by watching how long it takes to pass from maximum to maximum, and the distance of the globular in which it lies naturally follows. Originally there was some confusion because RR Lyrae variables were thought to follow the Cepheid period-luminosity law instead of having one of their own, but this misunderstanding has now been cleared up.

The brightest of the globular clusters visible in England is M.13, situated in Hercules (Map IX), which is fairly visible to the naked eye on a clear night. Binoculars will show it as a hazy patch, and a 3-inch will reveal stars near its edges, but to see it really well one needs an aperture of from 8 to 12 inches. Then, even the centre can be seen to consist of a myriad tiny points, and the sight is superb. Oddly enough, M.13 is unusually poor in RR Lyrae variables.

Globulars are much less common than the open clusters. Only about 100 are known, and most of these are faint, so that Messier listed only 28 of them. Unfortunately for us, the two finest globulars, Omega Centauri and 47 Tucanae, lie too far south to be visible in England.

Some of the Messier objects were different again; they looked like gas-patches, and were known as nebulae (Latin, "clouds"). For many years it was believed that the nebulae were merely star-clusters so far away that they could not be resolved with the telescopes available. This also applied to the curious "planetary nebulae", so called because they showed pale disks not unlike those of planets. But doubts began to creep in; some of the nebulae did not look in the least like clusters, and their real nature remained dubious.

---

*I drew up the catalogue quite casually one night, after an observing session, without any idea of publicizing it. On impulse I sent it to *Sky and Telescope* magazine; to my intense surprise it "caught on" and was well received – except by one rather sour amateur who was clearly annoyed because the idea was not his!

Fig. 16.3. M41, a prominent open cluster in Canis Major not far from Sirius. (Photograph by Phil Johnson 2001.)

By itself, the telescope could not solve the mystery, but the spectroscope came to the rescue. In 1864 Sir William Huggins, one of the great spectroscope pioneers, put the matter to the test by observing a planetary nebula in Draco. He half expected to see a somewhat confused effect due to the result of the combined spectra of thousands of stars, but instead he saw nothing but a single green line. At once he realized the truth. The light of the nebula was made up of one colour only, emitted by a luminous gas; the object was not a distant cluster at all, but something quite different (Figs 16.3 and 16.4).

Diffuse nebulae such as that in Orion's Sword had always been regarded as clouds of gas and dust in space; the planetaries too were found to be gaseous, but they are neither planets nor nebulae, so that their popular name could hardly be less apt.

A planetary nebula is a star at an advanced stage of evolution which has thrown off its outer layers and now consists of a small, very hot star surrounded by almost incredibly tenuous gas. (Our Sun will eventually go through the planetary nebula stage.) The best-known planetary is the Ring Nebula, M.57 Lyrae, close to Vega (Map VIII). It is easy to identify, since it lies between two fairly bright stars, the famous eclipsing binary Beta Lyrae and its neighbour Gamma. A 3-inch telescope will show it, but a larger aperture proves that it has the shape of a ring, not unlike a faintly luminous motor-car type, while the central star is only of the fifteenth magnitude. Some of the other planetaries are much less symmetrical.

Fig. 16.4. The globular cluster Omega Centauri, NGC 5139, much the brightest in the sky, but in the far south. (Photograph by Steve Quirk.)

Planetaries are most interesting objects, and it is a pity that most of them are so dim. However, there are plenty of diffuse nebulae. A few of them, notably M.42 – the Sword of Orion – can be seen with the naked eye. M.42 lies below the three stars of the Belt, as shown in Map IV, and cannot possibly be missed. It is one of the show-pieces of the sky, particularly as it contains the celebrated multiple star Theta Orionis, known commonly as the Trapezium because of the arrangement of its four brightest components. M.42 itself is 15 light-years across, and about 1,500 light-years away.

Many of these diffuse nebulae are within the range of a small telescope, but not all are of the same type. Some, such as the nebula contained in the Pleiades cluster, shine simply by reflecting the light of the intermingled stars, but others, including M.42 Orionis, show spectra which indicate that they are shining by themselves; the radiation from the mixed-in stars is affecting the gas and making it luminous. Like the planetaries, the diffuse nebulae are very rarefied, millions of times less dense than our own atmosphere. If you take a cubic inch of air and spread it out over a

Fig. 16.5. The Dumbbell Nebula, M27. (Photograph by Ron Arbour.)

cubic mile, you will end up with gas if about the same density as that in a nebula (Figs 16.5 and 16.6).

Spectroscopic work has led to our identifying many of the gases in nebulae. Hydrogen, helium and oxygen are all present, and there are also spectrum lines which were once thought to be due to a new element "nebulium", but which have disappointingly proved to be due merely to oxygen and nitrogen in unfamiliar states.

Diffuse nebulae shine because of the stars contained in them, and consequently a nebula that includes no convenient star will remain dark. Though it will therefore be invisible, it will make itself evident because it will blot out the stars or luminous gas behind it. It was once believed that the occasional well-defined starless patches were true "holes in the heavens", but it is now known that there is no basic difference between a nebula which shines and one which does not. The Coal Sack in the Southern Cross is much the most obvious of the dark nebulae; in Cygnus

Fig. 16.6. The Great Nebular in Orion, M42. (Photographed by Jerry Mulchin with his Meade LX200 10-inch at F/4 and ST8 CCD camera.)

Fig. 16.7. The Horse's Head Nebula in Orion, B33. (Photograph by Bill Patterson.)

there is the 'North America Nebula', so nicknamed because of its shape, which is quite easy to identify by using binoculars.

To observe nebulae a low magnification is to be preferred, except for planetaries. Quite a number of them are within the range of a small telescope, but they do tend to be decidedly elusive (Fig. 16.7).

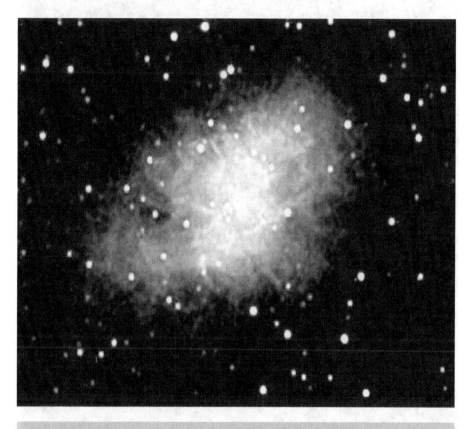

Fig. 16.8. The Crab Nebula, M1. (Photograph by Ron Arbour.)

Fig. 16.9. H × Perseus JRF. 10″ F6.3 Newtonian Refl ? prime focus 4.5 mins exposure TRIX 400 Developed in 019 for 4 minutes at 20° C. (Perseus)

Chapter 17

Galaxies

One of the glories of the night sky is the luminous band that is known to every-one as the Milky Way. It stretches right around the heavens, and on a clear moonless night it is a magnificent spectacle.

Galileo's first telescope, applied to the sky in the winter of 1609-10, led him to say that the Milky Way is made up of "an infinite number" of stars. This is an exagger-ation; the stars are not infinitely numerous, but there are about one hundred thousand million of them in our own system, together with a vast quantity of interstellar material.

Sweeping the Milky Way with binoculars or a low-power telescope will reveal so many stars that to count them by ordinary methods would be impossible. The belt is fairly well defined, and its stars seem to be bunched closely together, giving an impression of extreme over-crowding. But the universe is not a crowded place, and the stars in the Milky Way are no more packed than those in the rest of the sky. The luminous band itself is nothing more than a line-of-sight effect, due to the way in which our star-system or Galaxy is shaped.

A rough diagram of the Galaxy is given in Fig. 17.1. The stars are arranged in a form which bears some resemblance to two plates clapped together by their rims, with the Sun (S) well away from the centre. The dimensions of the "plate" are known with fair certainty, and the diameter (AB) proves to be 100,000 light-years, with the greatest breadth of it only one-fifth of this. We can now understand the reason for the Milky Way effect. When we look along SA or SB, we see many stars almost in the same line of sight, but when we look along SC or SD there are far fewer objects to be seen.

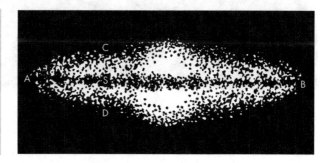

Fig. 17.1. Position of the Sun (S) in the Galaxy. Looking along the main plane (AB) results in the Milky Way effect.

Actually, it is not possible to see all the way from S to B. In the main plane of the Galaxy there is a great deal of obscuring material, both dust and gas, and starlight cannot penetrate it any more than a car's headlamps can penetrate a thick fog. The centre of nucleus of the Galaxy lies about 26,000 light-years in the direction of the rich star-clouds in Sagittarius (Map VIII), and here we have some glorious telescopic fields, but what lies beyond these fields can never be seen. Fortunately, the new science of radio astronomy has come to our rescue. Radio waves are not blocked by the interstellar matter, any more than a man's voice is blocked by fog, and we are at least learning more about the core of our stellar system.

Radio astronomy has also helped us to find out something about the structure of the Galaxy. It proves to be spiral, not unlike a vast Catherine-wheel, and the whole system is in rotation round its centre. The Sun takes about 225 million years to complete one circuit, so that it has been round only once since the far-off times when the Coal Measures were being laid down in the period of Earth history known to geologists as the Carboniferous. But though the Sun is moving round the centre in an almost circular orbit, other stars have paths of different types. It has now been established that there are two distinct "families" of stars, known generally as Populations I and II (Fig. 17.2).

Population I stars, such as the Sun, are found in the spiral arms of Galaxy. They are of various spectral types, but the most luminous of them are Blue Giants. On the other hand the senior members of Population II are Red Giants of vast size and relatively low temperature. Population II stars are found in the nucleus of the Galaxy, and also penetrate the vacant spaces between the spiral arms, while the stars of globular clusters also belong to this type. Since some Population II objects are revolving more slowly round the nucleus, and in more elliptical orbits, they seem to have high velocity with respect to ourselves, just as a slow-moving push-

Fig. 17.2. The Milky Way. This composite picture was taken at the Lund Observatory, Sweden. The two blobs below the main Milky Way system are the Magellanic Clouds.

bicyclist will seem to have "high velocity" relative to a stream of cars which is moving steadily as a group.

It is also interesting to note that in Population II areas, there is less of the cosmic obscuring matter which is such a nuisance to us. The areas have in fact been "swept clean", but in Population I regions the Blue Giants are always associated with clouds of dust and bright gaseous nebulae.

It is pointless to say much about methods of observing the Milky Way, except that a low power is to be preferred unless some particular object such as a faint nebula is to be examined. There are innumerable rich star-fields, particularly in the Cygnus area and in Sagittarius, and one never tires of sweeping about in these glorious regions, even though the chances of making a useful discovery are very small.

Two of the most striking of the objects in southern skies are the Clouds of Magellan, or Nubeculae. There are two of them, one much more conspicuous than the other, and it used to be thought that they were detached portions of the Milky Way; but they are in fact separate systems – satellites of our Galaxy. The large Cloud is 169,000 light-years away, and the Small Cloud considerably further – about 190,000 light-years. Both clouds are prominent naked-eye features, and contain objects of all kinds. European observers never cease to regret that they lie so far south in the sky (Map XV) (Fig. 17.3).

Vast though they are, the Clouds are far smaller than the system in which we live. Until recently it was indeed thought that the Milky Way was the largest of all galaxies, so that it had a special status in the universe, but this is far from being the case. There are millions of galaxies within range of our telescopes, and there is no longer any reason to suppose that our own system is of exceptional size.

There is still a tendency to refer to the galaxies as "spiral nebulae", but this is a bad term. Not all the galaxies are spiral, and certainly none of them is a nebula in the proper sense of the word, though they do contain nebulae of the same type as the Sword of Orion, as well as open clusters and globulars.

The nearest of the major galaxies, M.31, is easily found, since it can be seen with the naked eye. It lies in Andromeda, and is shown in Map VII. For many years it was thought to lie inside our own system, even though its spectrum showed that it was made up of stars and was not a luminous gas-cloud; but there were also suspicions that it might lie outside the Milky Way altogether. The riddle was solved in 1923, when Cepheids were discovered in the spiral arms. These Cepheids proved to be so remote that M.31 could no longer be regarded as a member of our Galaxy, and a first estimate of its distance gave a value of 900,000 light-years (Fig. 17.4).

Other Cepheids were found, and in 1944 a correction was made which reduced the distance of the galaxy to 750,000 light-years. All seemed to be well, but in 1952 stellar astronomers had a rude shock. It was found that the Cepheid scale was badly in error, because the difference between Population I and Population II Cepheids had not been realized; the result was that the whole distance-scale of the outer universe had to be doubled. Instead of lying at a mere 750,000 light-years, the Andromeda Galaxy was a million and a half light-years away. Further investigations have increased this still more, and the latest estimate is 2,900,000 light-years. The light now entering our eyes started on its journey towards us before the beginning of the last Ice Age.

Fig. 17.3. The Magellanic Clouds. (Photograph by Fred Espenak.)

It also followed that instead of being smaller than the Milky Way, the Andromeda Galaxy is larger, with a total mass at least $1\frac{1}{2}$ times as great. It too is in rotation; it too has its Population I stars (mostly in the arms) and Population II stars (mainly in the nucleus), as well as globulars, open clusters, gaseous nebulae and even two small satellite galaxies of the same status as our Clouds of Magellan. Novae have been observed, and in 1885, there was even a supernova which flared up to the sixth magnitude. At maximum it could only just be seen without a telescope, but had it appeared inside our own Galaxy it would probably have shone in our skies more brilliantly than Venus.

The Andromeda Galaxy is spiral, but as it is not face-on to us the spectacular Catherine-wheel effect is largely lost. I have always regarded it as rather a disappointing object in a small telescope, and in a 3-inch refractor it looks like a badly-defined patch of mistiness. Powerful telescopes are needed to show it really well, but the best results are obtained by means of photography.

Other galaxies are better presented, so that they appear as Catherine-wheels. In some cases, however, there is no spiral structure. A few galaxies (such as the Small

Fig. 17.4. The Andromeda Spiral, M31. This shows the spiral structure. (Photograph by Steve Barnes.)

Cloud of Magellan) are virtually formless, while others are elliptical or globular. It used to be thought that the different shapes of galaxies indicated different stages in evolution, but this plausible-sounding idea has now been rejected (Fig. 17.5).

Galaxies are apparently most numerous away from the Milky Way zone. This does not mean that there is anything lop-sided about their distribution; the effect is due purely to the obscuring matter near the galactic plane (AB in Fig. 17.1). There are groups of galaxies here and there, and the total number of known external systems is staggeringly great, though even the giant telescopes of today cannot show us more than a small part of the universe.

Spectra of galaxies are not particularly easy to study. Each is made up of the combined spectra of millions of bodies of all types and the result is bound to be much less clear than with a spectrum of a single star. However, one thing has become clear: nearly all the spectra of galaxies show a red shift, which indicates a velocity of recession.

Apart from the Andromeda Spiral, the fainter spiral in Triangulum, the two Nubeculae and more than twenty minor systems which make up our own "local group", all the galaxies are racing away from us, and the more distant they are the faster they go. We have now found galaxies which are over 12,000 million light-years away, and are receding at more than 90 per cent of the speed of light.

It was the American astronomer Edwin Hubble who proved that the entire universe is expanding, with every group of galaxies racing away from every other

Fig. 17.5. The Whirlpool Galaxy, M51, in Canes Venatici. This photograph brings out the spiral structure and also the satellite galaxy. (Photograph by Darin Fields.)

group. The red shifts do not indicate that our own particular area is in any way exceptional; the situation may be visualized by picturing a balloon filled with coloured gas – when the balloon is burst, the gas expands, each part of it receding from each other apart. The analogy is admittedly not very accurate, but it is the best that can be done.

In 1963 there was a new development. By then, many radio sources had been tracked down, and identified either with galaxies, supernova remnants or other known objects. However, some sources seemed to coincide in position with stars. One, in particular – the source known by its catalogue number of 3C-273 – coincided with what seemed like a rather faint bluish star, which had been recorded photographically often enough. The whole situation seemed peculiar, and when the radio astronomers asked the American optical astronomers to take a closer look at the spectrum of the "star", some amazing facts emerged. The main surprise was that the bluish object was not a star at all. It had a totally different kind of spectrum, and a tremendous red shift, which presumably meant that it was very remote and was receding very rapidly. This was the first-identified of the objects now known as quasi-stellar objects or, more commonly, as quasars.

Quasars are now known to be the cores of very active galaxies. They are immensely luminous, and very remote, but quasars need not concern us here, because only one is within the range of a small telescope – and when found, it looks exactly like a star. I observed it recently with my $12\frac{1}{2}$-inch reflector, and it was hard to credit that this tiny twinkling point of light was as powerful as so many thousands of millions of suns.

Chapter 18

Beginnings and Endings

How did the universe begin – and how will it end, if indeed it ends at all? These are two questions which cannot yet be answered. We have, of course, made progress. Not so long ago a leading churchman, Archbishop James Usher of Armagh, announced that the world came into existence at nine o'clock in the morning of 26 October, BC 4004 (I never found out whether he made due allowance for Summer Time). He based this estimate upon adding up the ages of the patriarchs, and making other calculations which were about as relevant as the flowers that bloom in the spring. We now know that the Earth is around 4,600 million years old; the Sun is older than that, and the universe older still.

According to modern theory, everything – space, matter, time – was created in a 'Big Bang' 13,700 million years ago. What happened before that? If the theory is correct, there was no 'before'; the Big Bang also marked the start of time.* The infant universe was inconceivably small and inconceivably hot, but it began to expand and cool. Galaxies condensed out of the original material; stars condensed out of galaxies, planets were born, life appeared at least on our Earth, and the universe developed into the form we know today. Accept the Big Bang, and we can work through a complete sequence, ending with you and me. What we cannot do is to explain the Big Bang itself; we simply do not have the mental equipment to understand it.

What of the future? Will the groups of galaxies continue to move away from each other, and will all the stars eventually die – or will the galaxies come together again in a new Big Bang? The latest research indicates that the expansion will never stop, and the universe will end up as cold and dead, but theories seem to change almost every week, and the jury is still out.

One day, perhaps, we will find out, but for now we have to accept our limitations. We are creatures of the present; the universe is spread out for our inspection, and everybody can play a part, from the astronomer who has the use of the most modern equipment down to the humble amateur who studies the Moon with a portable telescope set up in his back garden.

---

*The term "Big Bang" was used in a derisory sense by the great astronomer Fred Hoyle, who never believed in it.

Appendix 1

Planetary Data

Planet	Distance from Sun, in millions of miles			Sidereal Period	Synodic Period days	Axial Rotation (Equatorial)
	Max.	Mean	Min.			
MERCURY	43	36	29	88 days	115.9	58 d. 5 h.
VENUS	67.6	67.2	66.7	224.7 "	583.9	243 d.
EARTH	94.6	93.0	91.4	365 "	–	23 h. 56 m.
MARS	154.5	141.5	128.5	687 "	779.9	24 h. 37 m. 23 s.
JUPITER	506.8	483.3	459.8	11.86 years	398.9	9 h. 50 m. 30 s.
SATURN	937.6	886.1	834.6	29.46 "	378.1	10 h. 14 m.
URANUS	1,867	1,783	1,699	84.01 "	369.7	17 h. 14 m.
NEPTUNE	2,817	2,793	2,769	164.79 "	367.5	16 h. 7 m.
PLUTO	4,566	3,666	2,766	247.70 "	366.7	6 d. 9 h.

Planet	Diameter in miles (equatorial)	Apparent Diameter seconds of arc		Maximum Magn.	Axial Incl. Degrees	Mass Earth-I	Vol.
		Max.	Min.				
MERCURY	3030	12.9	4.5	–1.9	0	0.04	0.06
VENUS	7523	66.0	9.6	–4.4	178	0.83	0.88
EARTH	7,927	–	–	–	23.5	1	1
MARS	4,222	25.7	3.5	–2.8	24.0	0.11	0.15
JUPITER	89,424	50.1	30.4	–2.5	3.1	318	1,312
SATURN	74,914	20.9	15.0	–0.4	26.7	95	763
URANUS	31,770	3.7	3.1	+5.6	98	15	67
NEPTUNE	31,410	2.2	2.0	+7.7	29	17	57
PLUTO	1444	0.2	0.1	+14	122	0.002	0.007

Planetary Satellites of Magnitude 14.5 or Brighter

Name	Mean distance from centre of primary thousands of miles	Orbital Period d	h	m	Orbital Eccentricity e	Orbital Inclination, I degrees	Diameter, (Longest) miles	Magnitude
EARTH								
Moon	239.0	27	7	43	0.055	5.1	2160	−12.5
MARS								
Phobos	5.8	0	7	39	0.02	1.1	17	11.8
Deimos	14.6	1	6	18	0.003	1.8	9	12.8
JUPITER								
Io	262	1	18	28	0.004	0.04	2264	5.0
Europa	417	3	13	14	0.009	0.47	1945	5.3
Ganymede	666	7	3	43	0.002	0.21	3274	4.6
Callisto	1170	16	16	32	0.007	0.31	2981	5.6
SATURN								
Mimas	115	0	22	37	0.020	1.52	245	12.9
Enceladus	148	1	8	53	0.004	0.02	318	11.8
Tethys	183	1	21	18	0.000	1.09	663	10.3
Dione	235	2	17	41	0.000	0.02	696	10.4
Rhea	328	4	12	25	0.001	0.35	950	0.7
Titan	760	15	22	41	0.029	0.33	3201	8.3
Hyperion	920	21	6	38	0.104	0.43	224	14.2
Iapetus	2210	79	7	56	0.028	14.7	912	11 (max)
URANUS								
Ariel	119	2	12	29	0.001	0.03	720	14.1
Umbriel	166	4	3	28	0.004	0.13	727	14.8
Titania	272	8	16	56	0.001	0.014	981	14.0
Oberon	365	13	11	7	0.001	0.07	947	14.2
NEPTUNE								
Triton	220	5	21	3	0.0000	157.3	1681	13.6

Appendix 3

Minor Planet Data

The following list includes data for the first ten minor planets to be discovered. Objects with interesting orbits, such as the Trojans and the "Earth-grazers", are in general too faint to be seen with amateur-owned equipment.

Number	Name	Diameter, miles	Sidereal Period, years	Mean Dist. from Sun, millions of miles	Orbital Incl., deg.	min.	Max. Mag.
1	CERES	601	4.60	257.0	10	36	7.4
2	PALLAS	355	4.61	257.4	34	48	8.7
3	JUNO	155	4.36	247.8	13	00	8.0
4	VESTA	326	3.63	219.3	7	08	6.0
5	ASTRÆA	75	4.14	239.3	5	20	9.9
6	HEBE	127	3.78	225.2	14	45	8.5
7	IRIS	129	3.68	221.5	5	31	8.7
8	FLORA	90	3.27	204.4	5	54	9.0
9	METIS	30	3.69	221.7	5	36	8.3
10	HYGEIA	281	5.59	292.6	3	49	9.5

Appendix 4

Elongations and Transits of the Inferior Planets

Elongations of Mercury, 2006–2015

Eastern

2006	24 Feb, 20 June, 17 Oct
2007	7 Feb, 2 June, 29 Sept
2008	22 Jan, 14 May, 11 Sept
2009	4 Jan, 26 Apr, 24 Aug, 18 Dec
2010	8 Apr, 7 Aug, 1 Dec
2011	23 Mar, 20 July, 14 Nov
2012	5 Mar, 4 July, 26 Oct
2013	16 Feb, 12 June, 9 Oct
2014	31 Jan, 25 May, 21 Sept
2015	14 Jan, 7 May, 4 Sept, 29 Dec

Western

2006	8 Apr, 7 Aug, 23 Nov
2007	22 Mar, 26 July, 8 Nov
2008	3 Mar, 1 July, 22 Oct
2009	13 Feb, 13 July, 22 Oct
2010	27 Jan, 26 May, 19 Sept
2011	9 Jan, 7 May, 3 Sept, 23 Dec
2012	18 Apr, 16 Aug, 4 Dec
2013	31 Mar, 30 July, 18 Nov
2014	14 Mar, 12 July, 1 Nov
2015	24 Feb, 24 June, 16 Oct

There will be transits on 2006 Nov 8 and 2016 May 9.

Venus

E elongation	Inferior conjuction	W elongation	Superior conjunction
2007 June 9	2007 Aug 18	2007 Oct 28	2006 Oct 27
2009 Jan 14	2009 Mar 27	2009 June 5	2008 June 9
2010 Aug 20	2010 Oct 29	2011 Jan 8	2010 Jan 12
2012 May 27	2012 June 5	2012 Aug 15	2011 Aug 16
2013 Nov 1	2014 Jan 10	2014 Mar 22	2013 Mar 28
2015 June 6	2015 Aug 16	2015 Oct 26	2014 Oct 25

The maximum elongation of Venus during this period is 47°07′, but all elongations range from 45°23′ to 47°07′. At the superior conjunctions of 9 June 2008 Venus will actually be occulted by the Sun.

Appendix 5

Map of Mars

All Earth-based observations of Mars have now been superseded by the space-probe results, but it is still interesting to give a map showing the features with modest telescopes.

The map given here is based upon my own observations made during 1963. The opposition of 1963 was, of course, rather unfavourable, but at least Mars was well north of the Equator; the planet's Northern Hemisphere was tilted towards us. The polar cap was much in evidence, and there were various short-lived cloud phenomena here and there on the disk.

I do not claim that this map is of extreme precision; it is not intended to be. What I have done is to put in the features that I was able to observe personally, making the positions as accurate as possible, and taking care to omit everything about which I was not fully satisfied. I have retained the old (IAU) nomenclature.

Some of the Martian markings are very easy to observe. In 1963 I was able to see various dark features with a 3in refractor, without the slightest difficulty; most prominent of all are the Syrtis Major in the Southern Hemisphere and the Mare Acidalium in the Northern, though Sinus Sabaeus, Mare Tyrrhenum, and other dark areas are also clear. The more delicate objects require larger apertures; I doubt whether, in 1963, Solis Lacus could have been glimpsed with anything less than an $8\frac{1}{2}$ in reflector, though it is of course possible that a 6in would have shown it to observers with keener eyes than mine.

Obviously, not all the features shown here are visible at any one moment. The map was compiled from more than 50 separate drawings made at different times. Elusive features that were marked as "suspected only" have been omitted. Because the chart is drawn to a Mercator projection, the polar regions are not shown, but this does not much matter – the south pole was badly placed (indeed, the actual pole was tilted away from the Earth), and the north polar region was covered with its usual white cap, which shrank steadily as the Martian season progressed.

No doubt many observers using the same instruments would have seen more than I did. All I will claim is that my rough map, unlike some other published charts, does not show any features that are of dubious reality!

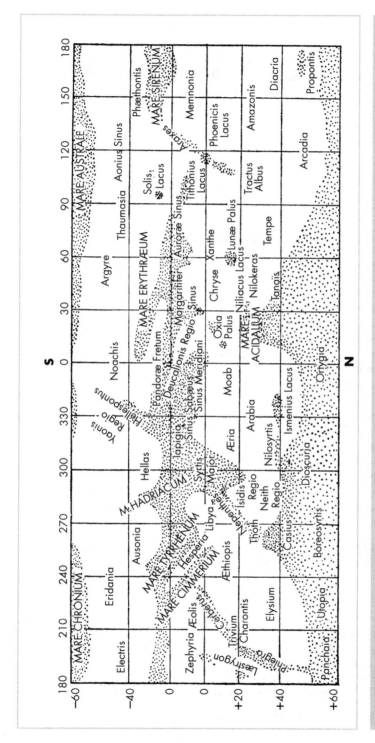

Map of Mars Drawn by Patrick Moore from personal observations in 1963 with his 12½ inch and 8½ inch reflectors. The projection is a Mercator, and the nomenclature follows that drawn up by the International Astronomical Union.

Jupiter: Transit Work

The diagram opposite shows Jupiter's main belts and zones:

SPR = South Polar Region
SSTZ = South South Temperate Zone
SSTB = South South Temperate Belt
STZ = South Temperate Zone
STB = South Temperate Belt
STrZ = South Tropical Zone
SEB = South Equatorial Belt (often double: SEBs = south component; SEBn = north component)
EZ = Equatorial Zone
E Band = Equatorial Band
NEB = North Equatorial Belt
NTrZ = North Tropical Zone
NTB = North Temperate Belt
NTZ = North Temperate Zone
NNTB = North North Temperate Belt
NNTZ = North North Temperate Zone
NPR = North Polar Region
RS = Great Red Spot

The following is a typical extract from my own observation diary:
1963 November 4, 12½-inch reflector. Conditions very variable.

GMT	Feature	Longitude System I	Longitude System II	Remarks
19.57	c. of white spot in STZ		182.1	× 360
20.06	f. of this white spot		187.5	
20.21	c. of white patch on the Equator	253.4		
20.22	p. of visible section of NTB		197.2	
20.30	f. of the white patch on the Equator	258.9		
20.37	f. of dark mass on N edge of NEB	263.1		

(c = centre, p = preceding, f = following)

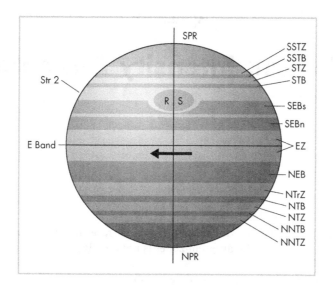

To work out the longitudes of the features, use the tables given in the B.A.A. *Handbook*, which give the longitude of the central meridian for various times.

Example. The 19.57 transit of the centre of the white spot in the STZ. From the tables: longitude of the central meridian (System II) for 16h on November 4 is 038.9. This is 3h 57m earlier than the time of the transit. Therefore, the longitude for 19.57 may be worked out from the second table in the *Handbook*:

Long at	16h. Nov. 4	=	038.9
	+ 3h	=	108.8
	+ 50m	=	30.2
	+ 7m	=	4.2
=	+ 3h 57m	=	182.1

If the calculated longitude works out at over 360, then subtract 360°.

It is important to use the correct System; System I is bounded by the N edge of the SEB and the S edge of the NEB, all the rest of the planet being System II. If the wrong tables are used, the results can be very peculiar indeed, since Systems I and II differ by many degrees.

Even more convenient tables are given in B.M. Peek's admirable book, *The Planet Jupiter*, and the example may then be worked as follows:

Long at	16h, November 4 (from the *Handbook*)	=	038.9
	+ 3h 57m (from Peek's tables)	=	143.2
			182.1

Saturn: Intensity Estimates

Definite features on the disk of Saturn are so rare that our knowledge of the rotation periods of the different zones is not merely so complete as in the case of Jupiter.

Valuable work can, however, be done in estimating the brightness of the different zones, as well as of the rings, as these are suspected of variation. The scale adopted is from a value of 0 (brilliant white) to 10 (black shadow). In general, Ring B is the brightest feature, and the outer part has a brightness of 1.

The easiest way of recording is to prepare a sketch (perhaps a rough one) of the globe and rings, and then merely jot down the numerical values upon the drawing itself. It is best to make each estimate twice; first, start from the darkest feature and work through to the lightest, then begin once more, this time with the lightest feature, which is almost always the outer part of Ring B. The following is an extract from my own notebook:

1956 May 21, oh. to oh. 20m. 12½-in. Refl X 460. Conditions good.

Ring B, outer	=	$1\frac{1}{2}$	N.E.B. intermediate zone	=	$5\frac{1}{2}$
Ring B, inner	=	$2\frac{1}{2}$	N. Equatorial Belt, N.		
Equatorial Zone	=	3	component	=	$6\frac{1}{2}$
N. Temperature Zone	=	$4\frac{1}{2}$	Encke's Division	=	7
N. Polar Region	=	5	Ring C	=	7
N. Temperate Belt	=	$5\frac{1}{2}$	Cassini's Division	=	$7\frac{1}{2}$
Ring A	=	$6\frac{1}{4}$	Shadow, Rings on Globe	=	8
N.E. Equatorial Belt, S			Shadow, Globe on Rings	=	$8\frac{1}{2}$
component	=	$6\frac{1}{2}$			

Of course, the different parts of the ring-system cannot be seen individually when the rings are edge-on to us, as in 1966.

Forthcoming Eclipses

		Solar	
Date	Mid-eclipse time GMT	Type	Area
2006 Mar. 29	10	Total	Atlantic, N. Africa, Turkey, Russia
2006 Sept. 22	12	Annular	Atlantic, S. Indian Ocean
2007 Mar. 19	03	Partial (88 pc)	N. America, Japan
2007 Sept. 11	13	Partial (75 pc)	S. America, Antarctic
2008 Feb. 7	04	Annular	S. Pacific, Antarctic
2008 Aug. 1	10	Total	Greenland, Siberia, China
2009 Jan. 26	08	Annular	Indian Ocean, Sri Lanka, Borneo

	Lunar	
Date	Mid-eclipse time GMT	Type
2006 Sept. 7	18.52	Partial (18 pc)
2007 Mar. 3	23.22	Total
2008 Feb. 21	03.27	Total
2008 Aug. 16	21.11	Partial (81 pc)

The Limiting Lunar Detail Visible with Different Apertures

The following information is based on work by E.A. Whitaker, formerly Director of the Lunar Section of the British Astronomical Association, and given in the Sectional journal, *The Moon* (Vol. 4, No. 2, page 42; December 1955). The table gives the approximate diameters of the smallest craters half-filled with shadow and of the narrowest black line certainly distinguishable. Perfect seeing conditions and first-class optical equipment are assumed.

Aperture of object glass in ins.	Smallest crater	Narrowest cleft
1	9 miles	0.5 mile
2	4.5 miles	0.25 mile
3	3 miles	0.16 mile
4	2.25 miles	220 yards
6	1.5 miles	150 yards
8	1.1 miles	110 yards
10	0.9 mile	90 yards
12	0.75 mile	70 yards
15	0.6 mile	60 yards
18	0.5 mile	50 yards
33	500 yards	30 yards

The smallest craterlet that I personally have recorded is probably that on the summit of a mountain peak near the crater Beer. The instrument used was the Meudon 33-inch, and the diameter of the summit depression cannot have been much more than 500 yards.

Appendix 10

The Lunar Maps

The following outline maps have been constructed from two photographs. The whole lunar surface is covered, but the method has two disadvantages. First, the formations near the eastern and western limbs are under high light and are consequently not well seen. Petavius, for instance, in the southeast, is really a majestic crater 100 miles across, and when anywhere near the terminator it is a magnificent object, but under this lighting it is hard to make out at all. Second, the photographs were taken when the Moon was at favourable libration for the east, so that the eastern limb regions are shown slightly better than the western. (Note that "east" and "west" are used in the astronautical sense, as described in Chapter 6, as the IAU has now ratified the change.)

These defects would be serious for a detailed map, but are not important for the present purpose. The observer may compare the map with the photograph given on the opposite page, and it will be easy to recognize the various formations. Once this has been done, serious work can be commenced; after a while, the observer will be able to identify the craters at a glance.

Only a few features are named on these charts; the remaining names will be found on more detailed maps.

The notes given here are, of course, extremely brief; they refer only to objects that are useful for "landmark" purposes, and one or two features of particular interest, such as Linné, the Alpine Valley, and the Straight Wall.

Fig. A.10.1. The Moon: eastern half.

Fig. A.10.2. The Moon: western half.

Names of the Lunar "Seas"

Latin	English
OCEANUS PROCELLARUM	OCEAN OF STORMS
MARE IMBRIUM	SEA OF SHOWERS
MARE FRIGORIS	SEA OF COLD
MARE HUMORUM	SEA OF HUMOURS
MARE VAPORUM	SEA OF VAPOURS
MARE FOECUNDITATIS	SEA OF FERTILITY
MARE TRANQUILLITATIS	SEA OF TRANQUILLITY
MARE SERENITATIS	SEA OF SERENITY
MARE CRISIUM	SEA OF CRISES
MARE HUMBOLDTIANUM	HUMBOLDT'S SEA
MARE SMYTHII	SMYTH'S SEA
MARE AUSTRALE	SOUTHERN SEA
MARE UNDARUM	SEA OF WAVES
MARE SPUMANS	FOAMING SEA
LACUS SOMNIORUM	LAKE OF DREAMS
PALUS SOMNI	MARSH OF DREAM
SINUS IRIDUM	BAY OF RAINBOWS
SINUS RORIS	BAY OF DEW
SINUS AESTUUM	BAY OF HEATS
SINUS MEDII	CENTRAL BAY
MARE MARGINIS	MARGINAL SEA

THE EASTERN HALF OF THE MOON (Fig. A10.1; Map I)

(1) *Northeastern Quadrant*

In this quadrant lie three major seas, the Maria Serenitatis, Tranquillitatis, and Crisium, with parts of the Mare Vaporum and the Mare Frigoris, as well as the northern part of the Mare Foecunditatis. Mare Serenitatis is the most conspicuous and is one of the best-defined of the lunar seas. On its surface are only two objects of importance, the 12-mile bright crater Bessel and the famous (or infamous) Linné, which used to be described as a deep crater, but is now seen in small telescopes as a mere white spot. A number of ridges cross the Mare Serenitatis. Mare Tranquillitatis is lighter in hue; between it and Serenitatis there is a strait upon which lies the magnificent 30-mile crater Plinius, which has two interior craterlets.

Of the mountain ranges, the most important are those bordering the Mare Serenitatis: the Haemus and the Caucasus Mountains, with peaks rising to 8,000 and 12,000 feet, respectively. Part of the Alps can also be seen, cut through by the strange Alpine Valley. This is an interesting formation, by far the most conspicuous of its type. South of Haemus, and north of the large crater Hipparchus, can be seen the two important clefts of the Mare Vaporum: that of Hyginus (which is basically a crater-chain) and Ariadaeus. Each can be seen with any small telescope when near the terminator.

Close to the eastern limb can be seen two very foreshortened seas, the Maria Smythii and Marginis, whereas the Mare Humboldtianum lies further north. These can only be well seen under favourable libration.

ARAGO. Diameter 18 miles. It lies on the Mare Tranquillitatis and has a low central elevation. Near it are several of the interesting swellings or "domes".

ARCHYTAS. A bright 21-mile crater on the north coast of the Mare Frigoris. It has a central peak.

ARISTILLUS. Diameter 35 miles, with walls rising to 11,000 feet above the floor. The walls are bright, and there is a central peak. Inside Aristillus are dark patches and streaks. It and Autolycus form a pair. Though it lies on the N.W. Quadrant, Aristillus is better shown in the next photograph.

ARISTOTELES. A prominent crater 52 miles in diameter. It and Eudoxus form a notable pair.

ATLAS. Forms a pair with Hercules; it lies north of the Mare Serenitatis. Diameter 55 miles. The walls are much terraced, rising to 11,000 feet. There is much detail on the floor.

AUTOLYCUS. The companion to Aristillus. Autolycus is 24 miles in diameter and 9,000 feet deep. It, too, is better shown on the next photograph.

BOSCOVICH. On the Mare Vaporum. It is low-walled, irregular formation, recognizable (like its companion Julius) by its very dark floor.

BÜRG. A 28-mile crater between Atlas and Aristoteles, with a large central peak on which is a summit craterlet. West of it lies an old plain traversed by numerous clefts.

CASSINI. On the fringe of the Alps. A curious broken formation, shallow, and 36 miles in diameter. It contains a prominent craterlet, A.

CLEOMEDES. A 78-mile crater near the Mare Crisium. It is broken in the west by a smaller but very deep crater, Tralles.

DIONYSIUS. A brilliant small crater near Sabine and Ritter.

ENDYMION. This 78-mile crater can always be recognized by the darkness of its floor. Patches on the interior seem to vary in hue and should be watched.

EUDOXUS. The companion to Aristoteles. It is 40 miles in diameter, and then 11,000 feet deep.

FIRMINICUS. Closely south of the Mare Crisium. It has a diameter of 35 miles, and can easily be identified by its dark floor.

GAUSS. A magnificent 100-mile crater, not well shown in the photograph, but very conspicuous when on the terminator.

GEMINUS. Diameter 55 miles. It lies near Cleomedes and has lofty walls, which are deeply terraced.

GODIN. Diameter 27 miles. It lies near Ariadaeus and Hyginus. Closely north lies Agrippa, which is slightly larger but somewhat less deep.

HERCULES. The companion of Atlas. It is 45 miles in diameter; the walls are much terraced and appear brilliant at times. Inside Hercules lies a large craterlet, A.

JULIUS CAESAR. A low-walled formation in the Mare Vaporum area. Owing to its dark floor, it is easy to recognize at any time.

MACROBIUS. A 42-mile crater near the Mare Crisium, with walls rising to 13,000 feet. There is low compound central mountain mass.

MANILIUS. A 25-mile crater on the Mare Vaporum, notable because of its brilliant walls.

MENELLAUS. Another brilliant crater; 20 miles in diameter, lying in the Haemus Mountains. Like Manilius, its brightness makes it easy to identify.

POSIDONIUS. A 62-mile plain on the border of the Mare Serenitatis. Adjoining it to the east is a smaller, squarish formation, Chacornac, and south is Le Monnier, one of the "bays" with a broken-down seaward wall.

PROCLUS. Closely west of the Mare Crisium. It is one of the most brilliant formations on the Moon and is the centre of a ray system. Diameter 18 miles.

SABINE AND RITTER. Two 18-mile craters on the west border of the Mare Tranquillitatis. Northeast of Ritter are two small equal craterlets. This area was photographed in detail by the U.S. probe Ranger VIII in 1965 and was again photographed from Apollo XI in 1969. The first lunar landing was made in the Mare Tranquillitatis, some distance east of Sabine and Ritter.

SCORESBY. A very distinct formation 36 miles across, near the North Pole. It is much the most conspicuous formation in its area and is thus very useful as a landmark.

TARUNTIUS. A 38-mile crater south of the Mare Crisium, with narrow walls and a low central hill. It is a "concentric crater", as it contains a complete inner ring.

(2) *South-East Quadrant*

This quadrant is occupied by rugged uplands, and large and small craters abound. The only major seas are the small, well-marked Mare Nectaris and most of the larger Mare Foecunditatis; the Mare Australe, near the limb, is much less well-defined. The only mountain summits of note are those very near the limb, some of which attain great altitudes. The so-called Altai Mountains are really in the nature of a scarp associated with the Mare Nectaris system.

ALBATEGNIUS. A magnificent walled plain near the centre of the disk; the companion of Hipparchus. Diameter 80 miles. The southwest wall is disturbed by a deep 20-mile crater, Klein.

CAPELLA. A 30-mile crater near Theophilus. It has a very long massive central mountain, topped by a summit craterlet; the floor of Capella is crossed by a deep valley. It has a shallower companion, Isidorus.

CUVIER. This forms an interesting group with Licetus and the irregular Heraclitus. Cuvier is 50 miles across and lies on the terminator in the photograph, not far from the top of the page.

FABRICIUS. A 55-mile crater, not well shown in the photograph owing to the high light. It has a companion of similar size, Metius. Fabricius interrupts the vast ruined plain Janssen.

GUTENBERG. This and its companion Goclenius lie on the highland between the Mare Nectaris and Foecunditatis. To the north lie some delicate clefts.

HIPPARCHUS. Well shown on the photograph, but it is low-walled and broken, so that it becomes obscure when away from the terminator. It is 84 miles in diameter and is the companion of Albategnius. Ptolemaeus lies closely west of it.

LANGRENUS. An 85-mile crater, with high walls and central mountain. It is a member of the great Eastern Chain, which extends from Furnerius in the south and includes Petavius, Vendelinus, Mare Crisium, Cleomedes, Geminus, and Endymion. It was in Langrenus that Dollfus detected "moving patches", thereby confirming the reality of Transient Lunar Phenomena (TLP).

MAUROLYCUS. A deep 75-mile crater, well shown in the photograph; in the far south of the Moon. West of it lies the larger plain Stöfler, which has a darkish floor.

MESSIER. This curious little crater lies on the Mare Foecunditatis. It and its companion, Messier A, show curious apparent changes in form and size. Extending west of them is a strange bright ray, rather like a comet's tail.

PETAVIUS. Not well shown on the photograph; but it is a magnificent object when better placed. Closely east of it is Palitzsch, which is generally described as a valley-like groove; but using very large telescopes, I found it to be a crater-chain. This was confirmed by orbiter photographs.

PICCOLOMINI. A 56-mile crater south of Fracastorious and the Mare Nectaris. It is deep and conspicuous, and lies at the eastern end of the Altai range.

RHEITA. A 42-mile crater. Associated with it is the famous Rheita Valley. This has been described as a "groove" and attributed to a falling meteorite; but is in fact a crater-chain, and no such explanation can be admitted. There is another similar formation not far off, associated with the crater. Reichenbach lies not far from the Mare Australe.

STEINHEIL. This 42-mile crater forms a pair of "Siamese twins" with its similar but shallower neighbour, Watt. Near the Mare Australe.

THEOPHILUS. The northern member of the grand chain of which Cyrillus and Catharina are the other members. Theophilus is 65 miles across and 18,000 feet deep; it is one of the most magnificent of the lunar craters. There is a lofty, complex central elevation.

VENDELINUS. One of the Eastern Chain. It is 100 miles across but is comparatively low-walled, and is conspicuous only when near the terminator.

VLACQ. One of a group of large ring-plains near Janssen, not far from the Mare Australe. It is 56 miles in diameter and 10,000 feet deep.

WERNER. A 45-mile crater, shadow-filled in the photograph. It forms a pair with its neighbour Aliacensis, and close by are three more pairs of formations: Apian-Playfair, Azophi-Abenezra, and Abulfeda-Almanon.

WILHELM HUMBOLDT. Not recognizable on the photograph, as it lies on the eastern limb near Petavius, but it is 120 miles across, with high walls and central mountain, and is magnificent just after full Moon.

THE WESTERN HALF OF THE MOON (Fig. A10.2; Map II)

(1) North-West Quadrant

This quadrant consists largely of "sea"; there is the magnificent Mare Imbrium, with a diameter of 700 miles, as well as parts of the even vaster but less well-defined Oceanus Procellarum, and most of the Mare Frigoris and the Sinus Roris. The chief mountains are the Apennines (certainly the most spectacular on the Moon) and the Jura Mountains, which form part of the Imbrian border. On the limb are the Hercynian Mountains. There are also the lower Carpathians, near Copernicus. Near full Moon, the most conspicuous objects are Copernicus and

Kepler, which are the centres of bright ray systems, and Plato, whose floor is so dark that it can never be mistaken, whereas Aristarchus is the most brilliant formation on the Moon. Eratosthenes, too, is a grand crater.

ANAXAGORAS. A 32-mile crater not far from the North Pole. It is the centre of a ray system, and is always distinct.

ARCHIMEDES. A 50-mile plain on the Mare Imbrium, with a darkish floor and rather low walls. It forms a superb group with Aristillus and Autolycus.

ARISTARCHUS. The brightest formation on the Moon. Associated with its companion, Herodotus, is a great winding valley. The whole area is particularly subject to Transient Lunar Phenomena (TLP) and should be systematically watched.

COPERNICUS. The great 56-mile ray-crater, described in the text.

SINUS IRIDUM. A glorious bay on the border of the Mare Imbrium. When the Sun is rising over it, the rays catch the bordering Jura Mountains, and the bay seems to stand out into the darkness like a handle of glittering jewels.

KEPLER. A 22-mile crater on the Oceanus Procellarum, centre of a very conspicuous system of bright rays. South of it is a crater of similar size, Encke, which is however shallower and is not associated with any bright rays.

OLBERS. A crater on the west limb. It lies north of Grimaldi, and is 40 miles in diameter. It is not identifiable on the photograph, because of the high light and unfavourable libration; but it is prominent when well placed, and is the centre of a ray system.

PHILOLAUS. A crater near the limb, 46 miles in diameter. It forms a pair with its neighbour Anaximenes. Reddish hues have been reported inside Philolais, perhaps indicating some unusual surface deposit.

PICO. A splendid 8,000-foot mountain on the Mare Imbrium, south of Plato, with at least three peaks. Some way southeast of it is Piton, which is also shown on the first photograph and has a summit craterlet.

PLATO. This regular, 60-mile formation has a dark floor, and is one of the most interesting features on the Moon. Inside it are some delicate craterlets that show baffling changes in visibility. Plato is always identifiable, and will well repay close and continuous attention.

PYTHAGORAS. A very deep crater 85 miles in diameter, not well shown in the photograph, but magnificent when well placed. There are numerous large formations in this area, but the whole region is very foreshortened as seen from Earth.

STRAIGHT RANGE. A peculiar range of peaks on the Mare Imbrium, near Plato. It is 40 miles long, and the highest mountains attain 6,000 feet.

TIMOCHARIS. A 23-mile crater on the Mare Imbrium, containing a central craterlet. It is the centre of a rather inconspicuous system of rays.

(2) South-West Quadrant

This quadrant is crammed with interesting features. In the northern part of it lie the well-marked Marc Humorum, part of the Oceanus Procellarum, and most of the vast Mare Nubium: the southern part is mainly rough upland. The chief mountain ranges are the curious low Riphaeans, on the Mare Nubium; the Percy Mountains, forming part of the border of the Mare Humorum; and the Dörfels,

Rook Mountains, Cordilleras, and D'Alemberts on the limb. It is now known that these ranges are associated with the Mare Orientale, which is never well seen from Earth; orbiter and Apollo pictures show it to be a vast, complex structure, unlike anything else on the Moon.

ALPHONSUS. The great crater close to Ptolemaeus. Dark patches may be seen on its floor. It was at Alphonsus that Kozyrev, in 1958, reported a visible outbreak of activity. The U.S. vehicle Ranger IX landed in Alphonsus in 1965.

BAILLY. Very obscure on the photograph; but it is almost 180 miles across, and on the Earth-turned hemisphere is thus the largest of the objects generally classed as "craters". It has been aptly described as a "field of ruins".

BILLY. A 30-mile crater S. of Grimaldi. It can be identified at any time because of its very dark floor; it is always distinct. It has a near neighbour, Hansteen, with a much lighter floor.

BIRT. A crater 11 miles in diameter, in the Mare Nubium, near the Straight Wall. It has walls that rise unusually high above the outer plain, and inside it are two of the strange radial bands.

BULLIALDUS. A splendid 39-mile crater on the Mare Nubium, with terraced walls and a central peak. This is one of the most perfect of the ring-plains.

CLAVIUS. Clavius is 145 miles across, with walls containing peaks 17,000 feet above the floor. Inside it can be seen a chain of craters, decreasing in size from east to west. When right on the terminator, Clavius can be identified with the naked eye.

CRÜGER. A low-walled crater near Grimaldi, 30 miles in diameter. It can be identified on the photograph by the darkness of its floor, which is rather similar to Billy's.

DOPPELMAYER. An interesting 40-mile bay on the Mare Humorum. The seaward wall can just be traced, and there is a much reduced central mountain.

EUCLIDES. Only 7 miles in diameter, but surrounded by a prominent bright nimbus, well shown on the photographs. It lies near the Riphaean Mountains.

FRA MAURO. One of a group of damaged ring-plains on the Mare Nubium. The other members of the group are Parry, Bonpland, and Guericke. The unlucky Apollo 13 astronauts were scheduled to land in this area, subsequently assigned to Apollo 14.

GASSENDI. A magnificent walled plain on the northern border of the Mare Humorum. It is 55 miles in diameter, and the floor contains a central mountain and numerous clefts. Reddish patches have been seen in and near Gassendi and are described in the text.

HIPPALUS. Another bay on the Mare Humorum, not unlike Doppelmayer. Near it are numerous prominent clefts, well seen in a small telescope, and there are also clefts on the floor. Near Hippalus is a small crater, Agatharchides A, in which I discovered two radial bands. These bands are useful test objects. I have seen them clearly with an aperture of 6 inches, but keener eyed observers should detect them with smaller instruments.

GRIMALDI. Identifiable at all times because of its floor, which is the darkest spot on the Moon. It lies close to the west limb. Patches on the floor show interesting variations in hue and should be watched. Grimaldi has low walls, and is

120 miles in diameter. Nearby is a smaller formation, Riccioli, 80 miles in diameter; it too has a very dark patch inside it.

LETRONNE. A bay of 70 miles in diameter lying on the shore of the Oceanus Procellarum not far from Gassendi. There is the wreck of a central elevation.

MAGINUS. A vast walled plain near Clavius and Tycho. It is very prominent when near the terminator, as in the photograph; but it becomes very obscure near full Moon.

MERCATOR. This and Campanus form a conspicuous pair of craters east of the Mare Humorum. Each is about 28 miles in diameter, and the only obvious difference between them is that Mercator has a darker floor.

MERSENIUS. A convex-floored 45-mile crater near Gassendi, associated with an interesting system of clefts.

MORETUS. Not well-known on the photograph, but it is a splendid crater 75 miles in diameter and 15,000 feet deep. The central mountain is the highest of its types on the Moon.

PITATUS. Described by Wilkins as being like a "lagoon". It lies on the south border of the Mare Nubium and has a dark floor and a low mountain near its centre. It is 150 miles in diameter. West of it is a smaller formation, Hesiodus, and from Hesiodus a prominent cleft runs towards Mercator and Campanus.

PTOLEMAEUS. More than 90 miles across; one of the most interesting formations on the Moon. It lies near the centre of the disk. Its floor is moderately dark. It is the northern member of a chain of three craters, the other two being Alphonsus and Arzachel. South of this chain lies another, made up of the three formations Purbach, Regiomontanus, and Walter.

SCHICKARD. A formation 134 miles in diameter. It can be identified on the photograph, near the southwest limb, because parts of its floor are darkish. Obscurations have been reported inside it, and it is well worth watching.

SIRSALIS. This and its "Siamese twin", A, lie near the dark-floored Crüger, not far from Grimaldi. Unfortunately they are not identifiable on the photograph. Sirsalis is associated with one of the most prominent clefts on the Moon.

STRAIGHT WALL. The celebrated fault in the Mare Nubium, near Birt. It is shown in the photograph as a white line, but casts considerable shadow before full Moon, when the illumination is from the reverse direction, so that it then appears as a dark line. Near it are numerous craterlets, some of them visible with very modest apertures. The Wall lies inside a large and obscure ring.

THEBIT. A 37-mile crater near the Straight Wall. It is interrupted by a small crater, which is in turn interrupted by a third. The group makes a useful test object for small apertures.

TYCHO. The great ray-crater, described in the text.

VITELLO. A 30-mile crater on the border of the Mare Humorum, with an inner but not quite concentric ring.

WARGENTIN. Most unfortunately, this is not identifiable on the photograph. It lies near Schickard, and is a 55-mile plateau, much the largest formation of its type on the Moon. Little detail can be seen in small telescopes; it is nevertheless worth observing. Near Wargentin is an interesting group of craters of which Phocylides is the largest member.

Some of the More Important Annual Meteor Showers

This list includes only a few of the many annual showers. The dates given for the beginnings and ends of the showers are only approximate.

Name	Beginning	End	Naked-eye star near radiant	Remarks
QUADRANTIDS	Jan. 3	Jan. 5	Beta Boötis	Usually a sharp maximum, Jan. 4.
LYRIDS	Apr. 19	Apr. 22	Nu Herculis	Moderate shower. Swift meteors.
ETA AQUARIDS	April 21	May 12	Eta Aquarii	Long paths; very swift.
DELTA AQUARIDS	July 15	Aug. 10	Delta Aquarii	Moderate shower.
PERSEIDS	July 27	Aug. 17	Eta Persei	A rich shower. Meteors very swift.
ORIONIDS	Oct. 15	Oct. 25	Nu Orionis	Moderate shower. Swift meteors.
LEONIDS	Nov. 14	Nov. 20	Zeta Leonis	Not usually a rich shower. Very swift meteors.
ANDROMEDIDS	Nov. 26	Dec. 4	Gamma Andromedae	Very slow meteors. Very weak shower.
GEMINIDS	Dec. 6	Dec. 19	Castor	Very rich shower.
URSIDS	Dec. 20	Dec. 22	Kocab	Rather weak.

The Constellations

In the following list, an asterisk indicates that the constellation was listed by Ptolemy; X, that much or all of the constellation is invisible in Europe. Zodiacal constellations are distinguished by the letter Z.

Constellations	English Names	Remarks	First Magnitude Star or Stars
Andromeda	Andromeda	*	
Antlia	The Air-Pump	X	
Apus	The Bird of Paradise	X	
Aquarius	The Water-Bearer	*Z	
Aquila	The Eagle	*	Altair
Ara	The Altar	X	
Aries	The Ram	*Z	
Auriga	The Charioteer	*	Capella
Boötes	The Herdsman	*	Arcturus
Caelum	The Sculptor's Tools	X	
Camelopardus	The Camelopard	–	
Cancer	The Crab	*Z	
Canes Venatici	The Hunting Dogs	–	
Canis Major	The Great Dog	*	Sirius
Canis Minor	The Little Dog	*	Procyon
Capricornus	The Sea-Goat	*Z	
Carina	The Keel	X	Canopus
Cassiopeia	Cassiopeia	*	
Centaurus Agena	The Centaur	X	Alpha Centauri
Cephus	Cephus	*	
Cetus	The Whale	*	
Chamaeleon	The Chameleon	X	
Circinus	The Compasses	X	
Columba	The Dove	–	
Coma Berenices	Berenice's Hair	–	
Corona Australis	The Southern Crown	X	

Corona Borealis	The Northern Crown	*	
Corvus	The Crow	–	
Crater	The Cup	–	
Crux Australis	The Southern Cross	X	Acrux, Beta Crucis
Cygnus	The Swan	*	Deneb
Delphinus	The Dolphin	*	
Dorado	The Swordfish	X	
Draco	The Dragon	*	
Equuleus	The Little Horse	*	
Eridanus	The River Eridanus	*X	Achernar (X)
Fornax	The Furnace	–	
Gemini	The Twins	*Z	Pollux
Grus	The Crane	X	
Hercules	Hercules	*	
Horologium	The Clock	X	
Hydra	The Sea-Serpent	*	
Hydrus	The Water-Snake	X	
Indus	The Indian	X	
Lacerta	The Lizard	–	
Leo	The Lion	*Z	Regulus
Leo Minor	The Little Lion	–	
Lepus	The Hare	–	
Libra	The Scales	*Z	
Lupus	The Wolf	X	
Lynx	The Lynx	–	
Lyra	The Harp	*	Vega
Mensa	The Table	X	
Microscopium	The Microscope	X	
Monoceros	The Unicorn	–	
Musca Australis	The Southern Fly	X	
Norma	The Rule	X	
Octans	The Octant	X	
Ophiuchus	The Serpent-Bearer	*	
Orion	Orion	*	Rigel, Betelgeux
Pavo	The Peacock	X	
Pegasus	The Flying Horse	*	
Perseus	Perseus	–	
Phoenix	Phoenix	X	
Pictor	The Painter	X	
Pisces	The Fishes	*Z	
Piscis Austrinus	The Southern Fish	*	Fomalhaut
Puppis	The Poop	*X	
Pyxis	The Compass	*	
Reticulum	The Net	X	
Sagitta	The Arrow	*	
Sagittarius	The Archer	*Z	
Scorpius	The Scorpion	*Z	Antares

Sculptor	The Sculptor	–	
Scutum	The Shield	–	
Serpens	The Serpent	*	
Sextans	The Sextant	–	
Taurus	The Bull	*Z	Aldebaran
Telescopium	The Telescope	X	
Triangulum	The Triangle	*	
Triangulum Australe	The Southern Triangle	X	
Tucana	The Toucan	X	
Ursa Major	The Great Bear	*	
Ursa Minor	The Little Bear	*	
Vela	The Sails	*X	
Virgo	The Virgin	*Z	Spica
Volans	The Flying-Fish	X	
Vulpecula	The Fox	–	

---

*Some of the original names have been abbreviated; for instance, "Reticulum Rhomboidalis" (the Rhomboidal Net) is simply "Reticulum". A few constellations have alternative names; Ophiuchus may be called "Serpentarius".

Proper Names of Stars

Some of the stars have been given proper names. Most of these have now fallen into disuse, but because they are still produced occasionally, the observer may find it useful to have a list. The names listed here are by no means all that have been given, but include the more important examples.

A few stars have more than one name (Eta Ursae Majoris can be "Benetnasch" as well as "Alkaid"), and some names can be spelled in more than one way (Betelgeux can be "Betelgeuse" or "Betelgeuze"). It is clearly pointless to give all these variations.

Constellation	Greek letter	Name
Andromeda	Alpha	Alpheratz
	Beta	Mirach
	Gamma	Almaak
	Xi	Adhil
Aquarius	Alpha	Sadalmelik
	Beta	Sadalsuud
	Gamma	Sadachiba
	Delta	Scheat
	Epsilon	Albali
	Theta	Ancha
Aquila	Alpha	Altair
	Beta	Alshain
	Gamma	Tarazed
	Zeta	Dheneb
	Kappa	Situla
	Lambda	Althalimain
Ara	Alpha	Choo
Argo Navis	Zeta	Suhail Hadar
	Iota	Tureis
	Kappa	Markeb
	Lambda	Al Suhail Al Wazn
	Xi	Asmidiske
	Rho	Taurais

Aries	Alpha	Hamal
	Beta	Sheratan
	Gamma	Mesartim
	Delta	Boteïn
Auriga	Alpha	Capella
	Beta	Menkarlina
	Zeta	Sadatoni
	Iota	Hassaleh
Boötes	Alpha	Arcturus
	Beta	Nekkar
	Gamma	Seginus
	Epsilon	Izar
	Eta	Saak
	h	Merga
	Mu	Alkalurops
Cancer	Alpha	Acubens
	Gamma	Asellus Borealis
	Delta	Asellus Australis
	Zeta	Tegmine
Canes Venatici	Alpha	Cor Caroli
	Beta	Chara
	Schj 152	La Superba
Canis Major	Alpha	Sirius
	Beta	Mirzam
	Gamma	Muliphen
	Delta	Wezea
	Epsilon	Adara
	Zeta	Phurad
	Eta	Aludra
Canis Minor	Alpha	Procyon
	Beta	Gomeisa
Capricornus	Alpha	Al Giedi
	Beta	Dabih
	Gamma	Nashira
	Delta	Deneb al Giedi
	Nu	Alshat
Carina	Alpha	Canopus
	Beta	Miaplacidus
	Epsilon	Avior
	Iota	Tureis
Cassiopeia	Alpha	Shedir
	Beta	Chaph
	Gamma	Tsih
	Delta	Ruchbah
	Epsilon	Segin
	Eta	Achird
	Theta	Marfak

Centaurus	Alpha	Al Rijil* or Toliman
	Beta	Agena
	Gamma	Menkent
Cepheus	Alpha	Alderamin
	Beta	Alphirk
	Gamma	Alrai
	Xi	Kurdah
Cetus	Alpha	Menkar
	Beta	Diphda
	Gamma	Alkaffaljidhina
	Zeta	Baten Kaitos
	Iota	Deneb Kaitos Shemali
	Omicron	Mira
Columba	Alpha	Phakt
	Beta	Wezn
Coma Berenices	Alpha	Diadem
Corona Borealis	Alpha	Alphekka
	Beta	Nusakan
Corvus	Alpha	Alkhiba
	Beta	Kraz
	Gamma	Minkar
	Delta	Algorel
Crater	Alpha	Alkes
Crux Australis	Alpha	Acrux
	Beta	Minosa+
Cygnus	Alpha	Deneb
	Beta	Albireo
	Gamma	Sadr
	Epsilon	Gienah
	Pi	Azelfafage
Delphinus	Alpha	Svalocin
	Beta	Rotanev
Draco	Alpha	Thuban
	Beta	Alwaid
	Gamma	Etamin
	Delta	Taïs
	Epsilon	Tyl
	Zeta	Aldhibah
	Eta	Aldhibain
	Iota	Edasich
	Lambda	Giansar
	Mu	Alrakis

---

*The proper name for Alpha Centauri is not generally used, except by navigators, who refer to it as "Rigel Kent".

+As in the case of Alpha Centauri, the proper name for Beta Crucis seems to be regarded as "unofficial", and is not generally used.

	Xi	Juza
	Psi	Dziban
Equuleus	Alpha	Kitalpha
Eridanus	Alpha	Achernar
	Beta	Kursa
	Gamma	Zaurak
	Delta	Rana
	Zeta	Zibal
	Eta	Azha
	Theta	Acamar
	Omicron 1	Beid
	Omicron 2	Keid
	Tau	Angetenar
	53	Sceptrum
Gemini	Alpha	Castor
	Beta	Pollux
	Gamma	Alhena
	Delta	Wasat
	Epsilon	Mebsuta
	Zeta	Mekbuda
	Eta	Propus
	Mu	Tejat
	Xi	Alzirr
Grus	Alpha	Alnair
	Gamma	Al Dhanab
Hercules	Alpha	Rasalgethi
	Beta	Kornephoros
	Delta	Sarin
	Zeta	Rutilicus
	Kappa	Marsik
	Lambda	Masym
	Omega	Cujam
Hydra	Alpha	Alphard
Leo	Alpha	Regulus
	Beta	Denebola
	Gamma	Algieba
	Delta	Zosma
	Epsilon	Asad Australis
	Zeta	Adhafera
	Theta	Chort
	Lambda	Alterf
	Mu	Rassalas
	Omicron	Subra
Leo Minor	46	Praecipua
Lepus	Alpha	Arneb
	Beta	Nihal

Libra	Alpha	Zubenelgenubi
	Beta	Zubenelchemali
	Gamma	Zubenelhakrabi
	Sigma	Zubenalgubi
Lupus	Alpha	Men
	Beta	Ke Kouan
Lyra	Alpha	Vega
	Beta	Sheliak
	Gamma	Sulaphat
	Eta	Aladfar
	Mu	Al Athfar
Ophiuchus	Alpha	Rasalhague
	Beta	Cheleb
	Delta	Yed Prior
	Epsilon	Yed Post
	Zeta	Han
	Eta	Sabik
Orion	Alpha	Betelgeux
	Beta	Rigel
	Gamma	Bellatrix
	Delta	Mintaka
	Epsilon	Alnilam
	Zeta	Alnitak
	Eta	Algjebbah
	Kappa	Saiph
	Lambda	Heka
	Upsilon	Thabit
Pegasus	Alpha	Markab
	Beta	Scheat
	Gamma	Algenib
	Epsilon	Enif
	Zeta	Homan
	Eta	Matar
	Theta	Biham
	Mu	Sadalbari
Perseus	Alpha	Mirphak
	Beta	Algol
	Zeta	Atik
	Kappa	Misam
	Xi	Menkib
	Omicron	Ati
	Tau	Kerb
	Upsilon	Nembus
Phoenix	Alpha	Ankaa
Pisces	Alpha	Kaïtain
Piscis Australis	Alpha	Fomalhaut

Puppis	Zeta	Suhail Hadar
	Xi	Asmidiske
	Rho	Turais
Sagittarius	Alpha	Rukbat
	Beta	Arkab
	Gamma	Alnasr
	Delta	Kaus Meridionalis
	Epsilon	Kaus Australis
	Zeta	Ascella
	Lambda	Kaus Borealis
	Pi	Albaldah
	Sigma	Nunki
Scorpius	Alpha	Antares
	Beta	Graffias
	Gamma (= Sigma Librae)	Zubenalgubi
	Delta	Dschubba
	Epsilon	Wei
	Theta	Sargas
	Kappa	Girtab
	Lambda	Shaula
	Nu	Jabbah
	Sigma	Alniyat
	Upsilon	Lesath
	Omega	Jabhat al Akrab
Serpens	Alpha	Unukalhai
	Theta	Alya
Taurus	Alpha	Aldebaran
	Beta	Alnath
	Gamma	Hyadum Primus
	Epsilon	Ain
	Eta	Alcyone
	17	Electra
	19	Taygete
	20	Maia
	21	Asterope
	23	Merope
	27	Atlas
	28	Pleione
Triangulum	Alpha	Rasalmothallah
Triangulum Australe	Alpha	Atria
Ursa Major	Alpha	Dubhe
	Beta	Merak
	Gamma	Phad
	Delta	Megrez
	Epsilon	Alioth
	Zeta	Mizar
	Eta	Alkaid

	Iota	Talita
	Lambda	Tania Borealis
	Mu	Tania Australis
	Nu	Alula Borealis
	Xi	Alula Australis
	Omicron	Muscida
	Pi	Ta Tsun
	Chi	Alkafzah
	80	Alcor
Ursa Minor	Alpha	Polaris
	Beta	Kocab
	Gamma	Pherkad Major
	Delta	Yildun
	Zeta	Alifa
	Eta	Alasco
Vela	Gamma	Regor
	Kappa	Markeb
	Lambda	Al Suhail al Wazn
Virgo	Alpha	Spica
	Beta	Zawijah
	Gamma	Arich
	Delta	Minelauva
	Epsilon	Vindemiatrix
	Eta	Zaniah
	Iota	Syrma

Stars of the First Magnitude

The 21 brightest stars are recognized as being of the first magnitude. The values given here are the latest determinations and differ somewhat from earlier values.

Star	Name	Mag.	Spectrum	Distance, lt-yrs	Luminosity, Sun 1
Alpha Canis Majoris	Sirius	−1.44	A1	8.6	26
Alpha Carinae	Canopus	−0.62	F0	310	15,000
Alpha Centauri		−0.27	G2+K1	4.4	1.7 + 0.45
Alpha Boötis	Arcturus	−0.05	K2	37	115
Alpha Lyrae	Vega	0.03	A0	25	52
Alpha Aurigae	Capella	0.08	G8+G0	42	90 + 70
Beta Orionis	Rigel	0.18	B8	770	40,000
Alpha Canis Minoris	Procyon	0.40	F5	11	8
Alpha Eridani	Achernar	0.45	B5	144	1000
Alpha Orionis	Betalgeux	0.45	M2	430	11,000
Beta Centauri	Agena	0.61	B1	530	10,000
Alpha Aquilae	Altair	0.76	A7	16.8	10
Alpha Crucis	Acrux	0.77	B0+B1	320	3200 = 2000
Alpha Tauri	Aldebaran	0.87	K5	65	140
Alpha Virginis	Spica	0.98	B1+B2	260	2200
Alpha Scorpii	Antares	1.08	M1	600	9000
Beta Geminorum	Pollux	1.16	K0	35	33
Alpha Piscis Australis	Fomalhaut	1.17	A3	25	14
Alpha Cygni	Deneb	1.25	A2	1800	60,000
Beta Crucis	Mimosa	1.25	B0	300	260,000
Alpha Leonis	Regulus	1.36	B7	78	125

Next in order of brightness come Epsilon Canis Majoris (Adhara) 1.50, Alpha Geminorum (Castor) 1.58, Gamma Orionis 1.59 and Lambda Scorpii (Shaula) 1.62.

Appendix 15

Standard Stars for Each Magnitude

It may be helpful to learn the magnitudes of a few standard stars for each magnitude, and the following are suitable. The first-magnitude stars are listed separately, from Sirius (−1.44) to Regulus (+1.36), though Regulus is, of course, nearer $1\frac{1}{2}$ than 1.

Approximate magnitude	Star	Exact magnitude
$1\frac{1}{2}$	Epsilon Canis Majoris	1.50
	Alpha Geminorum (Castor)	1.58
	Lambda Scorpii	1.60
	Gamma Orionis	1.64
2	Alpha Arietis	2.00
	Beta Ursae Minoris (Kocab)	2.04
	Kappa Orionis	2.06
	Alpha Andromedae (Alpheratz)	2.06
$2\frac{1}{2}$	Gamma Ursae Majoris (Phad)	2.44
	Epsilon Cygni	2.46
	Alpha Pegasi	2.50
	Delta Leonis	2.57
3	Zeta Aquilae	2.99
	Gamma Boötis	3.05
	Delta Draconis	3.06
	Zeta Tauri	3.07
$3\frac{1}{2}$	Alpha Trianguli	3.45
	Zeta Leonis	3.46
	Beta Boötis	3.48
	Epsilon Tauri	3.54
4	Beta Aquilae	3.90
	Gamma Coronae Borealis	3.93

	Delta Ceti	4.04
	Delta Cancri	4.17
$4\frac{1}{2}$	Nu Andromedae	4.42
	Delta Ursae Minoris	4.44
	Nu Cephei	4.46
5	Rho Ursae Majoris	4.99
	Eta Ursae Minoris	5.04
	Delta Trianguli	5.07
	Zeta Canis Minoris	5.11
$5\frac{1}{2}$	Theta Ursae Minoris	5.33
	Rho Coronae Borealis	5.43
	Epsilon Trianguli	5.44

The Greek Alphabet

α	Alpha		ν	Nu
β	Beta		χ	Xi
γ	Gamma		o	Omicron
δ	Delta		π	Pi
ϵ	Epsilon		ρ	Rho
ζ	Zeta		σ	Sigma
η	Eta		τ	Tau
ϑ	Theta		θ	Upsilon
ι	Iota		ξ	Phi
κ	Kappa		ψ	Chi
λ	Lambda		υ	Psi
μ	Mu		φ	Omega

Stellar Spectra

Type	Surface temp. (degrees C)	Colour	Typical star	Remarks
W	36,000+	Greenish white	Gamma Velorum, WC7	Wolf-Rayet. Many bright lines; helium prominent
O	36,000+	Greenish white	Zeta Puppis, O5	Wolf-Rayet. Helium prominent
B	28,600	Bluish	Spica, B1	Helium prominent
A	10,700	White	Sirius, A1	Hydrogen lines prominent
F	7,500	Yellowish	Beta Cassiopeiae, F2	Calcium lines prominent
G (giant)	5,200	Yellow	Epsilon Leonis, G0	{Metallic lines very numerous
G (dwarf)	6,000	Yellow	Sun, G2	
K (giant)	4,230	Orange	Arcturus, K2	{Hydrocarbon bands appear
K (dwarf)	4,910	Orange	Epsilon Eridani, K2	
M (giant)	3,400	Orange-red	Betelgeux, M2	{Broad titanium oxide and calcium bands or flutings
M (dwarf)	3,400	Orange-red	Wolf 359, M6	
R	2,300	Orange-red	U Cygni	Carbon bands
N	2,600	Red	S Cephei, Ne	Carbon bands. Reddest of all stars
S	2,600	Red	R Andromedae	Some zirconium oxide bands. Mostly long-period variables

Types R and N are often combined as Type C. The coolest stars are now given as Types L and T. A separate class, Q, has been reserved for novae.

Limiting Magnitudes and Separations for Various Apertures

It is extremely difficult to give definite values for limiting magnitudes, because so much must depend upon individual observers. The following table must be regarded as approximate only. The third column refers to stars of equal brilliancy and of about the sixth magnitude. Where the components are unequal, the double will naturally be a more difficult object, particularly if one star is much brighter than the other.

Aperture of object glass in inches	Faintest magnitude	Smallest separation seconds of arc
2	10.5	2.5
3	11.4	1.8
4	12.0	1.3
5	12.5	1.0
6	12.9	0.8
7	13.2	0.7
8	13.5	0.6
10	14.0	0.5
12	14.4	0.4
15	14.9	0.3

Appendix 19

Angular Measure

It may be useful to give the angular distances between some selected stars, as this will be of use to those who are not used to angular measurement. The distance all around the horizon is of course 360 degrees, and from the zenith to the horizon 90 degrees; the Sun and Moon have angular diameters of about 0.5 degrees, which is the same as that of an old halfpenny (1 inch) held at a distance of 9 feet from the eye.

Degrees (approx.)	Stars
60	Polaris to Pollux: Alpha Ursae Majoris to Beta Cassiopeiae
50	Sirius to Castor: Polaris to Vega
45	Polaris to Deneb: Spica to Antares
40	Capella to Betelgeux: Castor to Regulus
35	Vega to Altair: Capella to Pollux
30	Polaris to Beta Cassiopeiae: Aldebaran to Capella
25	Sirius to Procyon: Vega to Deneb
20	Betelgeux to Rigel: Procyon to Pollux
15	Alpha Andromedae to Beta Andromedae: Alpha Centauri to Acrux
10	Betelgeux to Delta Orionis: Acrix to Agena
5	Alpha Ursae Majoris to Beta Ursae Majoris
$4\frac{1}{2}$	Castor to Pollux: Alpha Centauri to Agena
3	Beta Scorpii to Delta Scorpii
$2\frac{1}{2}$	Altair to Beta Aquilae
2	Altair to Gamma: Beta Lyrae to Gamma Lyrae
$1\frac{1}{2}$	Beta Arietis to Gamma Arietis
1	Atlas to Electra (Pleiades)

To find the diameter of a telescopic field, select some star very near the celestial equator (such as Delta Orionis or Zeta Virginis) and allow it to drift through the field. This time in minutes, multiplied by 15, will give the angular diameter of the field in minutes and seconds of arc. For instance, if Delta Orionis takes 1 minute 3 seconds to pass through the field, the diameter is 1 minute 3 seconds × 15, or 15′ 45″ of arc.

Appendix 20

Test Double Stars

The following list is only approximate, as again so much depends upon the observer as well as upon the precise conditions, but it may be useful as a rough guide.

Apertures of O.G. inches	Star	Magnitudes		Separation secs. of arc	Position angle degree
1	Acrux	1.4,	1.9	4.7	119
	Alpha Herculis	var.,	5.4	4.5	109
2	Rigel	0.1,	6.7	9.5	202
	Gamma Leonis	2.3,	3.5	4.3	122
	Epsilon Boötis	2.4,	5.0	2.8	334
3	Polaris	2.0,	8.9	18.3	217
	Theta Virginis	4.0,	9.0	7.2	343
4	Theta Aurigae	2.7,	7.1	2.8	332
	Eta Orionis	3.8,	4.8	1.4	079
	Delta Cygni	2.9,	6.4	2.1	240
	Iota Ursae Majoris	3.1,	10.8	7.4	002
5	Zeta Boötis	4.6,	4.6	1.2	309
	Omega Leonis	5.9,	6.7	1.0	129
8	Lambda Cassiopeiae	5.5,	5.8	0.6	180
	*Gamma2 Andromedae	5.4,	6.6	0.7	109
9	Eta Coronae Borealis	5.7,	5.9	0.7	069

*Gamma2 Andromedae is the smaller component of the easy double Gamma Andromedae.

Appendix 21

Extinction

When estimating the brightness of a naked-eye variable, care must be taken to allow for atmospheric dimming. The closer a star is to the horizon, the more of its light will be lost. The following table gives the amount of dimming for various altitudes above the horizon. Above an altitude of 45 degrees, extinction can be neglected for all practical purposes.

Altitude degrees	Dimming in magnitudes
1	3
2	2.5
4	2
10	1
13	0.8
15	0.7
17	0.6
21	0.4
26	0.3
32	0.2
43	0.1

Bright Novae

This lists all novae since 1572 that have become brighter than magnitude 6.0.

Year	Star	Max magnitude	Discoverer
1670	CK Vulpeculae	3	Anthelm
1848	V 841 Ophiuchi	4	Hind
1876	Q Cygni	3	Schmidt
1891	T Aurigae	4.2	Anderson
1898	V 1059 Sagittarii	4.9	Fleming
1901	GK Persei	0.0	Anderson
1903	DM Geminorum	5.0	Turner
1910	DI Lacertae	4.6	Espin
1912	DN Geminorum	3.3	Enebo
1918	V 603 Aquilae	−1.1	Bower
1918	GI Monocerotis	5.7	Wold
1920	V 476 Cygni	2.0	Denning
1925	RR Pictoris	1.1	Watson
1934	DQ Herculis	1.2	Prentice
1936	V 368 Aquilae	5.0	Tamm
1936	CP Lacertae	1.9	Gomi
1936	V 630 Sagittarii	4.6	Okabayasi
1939	BT Monocerotis	4.3	Whipple and Wachmann
1942	CP Puppis	0.4	Dawson
1960	V 446 Herculis	5.0	Hassell
1963	V 533 Herculis	3.2	Dahlgren and Pelter
1967	HR Delphini	3.7	Alcock
1968	LV Vulpeculae	4.9	Alcock
1970	FH Serpentis	4.4	Honda
1975	V 1500 Cygni	1.8	Honda
1984	QU Vulpeculae	5.6	Collins
1986	V 842 Centauri	4.6	McNaught

Year	Star	Max. magnitude	Discoverer
1986	V 842 Centauri	4.6	McNaught
1991	V 838 Herculis	5.0	Alcock
1992	V 1974 Cygni	4.3	Collins
1993	V 705 Cassiopeia	5.4	Kanatsu
1999	V 382 Velorum	2.5	Williams and Gilmore
1999	V1494 Aquilae	3.6	Pereira

The recurrent nova T Coronae (the "Blaze Star"), usually of about magnitude 10, flared up to the second magnitude in 1866 and again in 1946. Watch out for a repeat performance around 2026!

Appendix 23

Messier's Catalogue

Messier's famous catalogue of nebular objects includes most of the brightest nebulae and clusters visible in England. It is therefore useful to give his list, as most of the objects can be found by means of the star maps in Appendix 25 and can be picked up by means of small telescopes.

Number	Constellation	Type	Magnitude	Remarks
1	Taurus	Wreck of supernova	8.4	Crab Nebula (radio source)
2	Aquarius	Globular	6.3	
3	Canes Venatici	Globular	6.4	
4	Scorpius	Globular	6.4	
5	Serpens	Globular	6.2	
6	Scorpio	Open cluster	5.3	
7	Scorpius	Open cluster	4.0	
8	Sagittarius	Nebula	6.0	Lagoon Nebula
9	Ophiuchus	Globular	7.3	
10	Ophiuchus	Globular	6.7	
11	Scutum	Open cluster	6.3	Wild Duck Cluster
12	Ophiuchus	Globular	6.6	
13	Hercules	Globular	5.7	Great globular cluster
14	Ophiuchus	Globular	7.7	
15	Pegasus	Globular	6.0	
16	Serpens	Nebula and embedded cluster	6.4	
17	Sagittarius	Nebula	7.0	Omega or Horseshoe Nebula
18	Sagittarius	Open cluster	7.5	
19	Ophiuchius	Globular	6.6	
20	Sagittarius	Nebula	9.0	Trifid Nebula
21	Sagittarius	Open cluster	6.5	
22	Sagittarius	Globular	5.9	
23	Sagittarius	Open cluster	6.9	

Number	Constellation	Type	Magnitude	Remarks
24	Sagittarius	Open cluster	4.6	
25	Sagittarius	Open cluster	6.5	
26	Scutum	Open cluster	9.3	
27	Vulpecula	Planetary	7.6	Dumbbell Nebula
28	Sagittarius	Globular	7.3	
29	Cygnus	Open cluster	7.1	
30	Capricornus	Globular	8.4	
31	Andromeda	Spiral galaxy	4.8	Great Galaxy
32	Andromeda	Elliptical galaxy	8.7	Satellite of M.31
33	Triangulum	Spiral galaxy	6.7	Triangulum Spiral
34	Perseus	Open cluster	5.5	
35	Gemini	Open cluster	5.3	
36	Auriga	Open cluster	6.3	
37	Auriga	Open cluster	6.2	
38	Auriga	Open cluster	7.4	
39	Cygnus	Open cluster	5.2	
41	Canis Major	Open cluster	4.6	
42	Orion	Nebula	4±	Great Nebula in Orion
43	Orion	Nebula	9±	Part of Orion Nebula
44	Cancer	Open cluster	3.7	Praesepe
45	Taurus	Open cluster	–	Pleiades
46	Puppis	Open cluster	6.0	
49	Virgo	Elliptical galaxy	8.6	
50	Monoceros	Open cluster	6.3	
51	Canes Venatici	Spiral galaxy	8.1	Whirlpool Galaxy
52	Cassiopeia	Open cluster	7.3	
53	Coma Berenices	Globular	7.6	
54	Sagittarius	Globular	7.3	
55	Sagittarius	Globular	7.6	
56	Lyra	Globular	8.2	
57	Lyra	Planetary	9.3	Ring Nebula
58	Virgo	Spiral galaxy	8.2	
59	Virgo	Elliptical galaxy	9.3	
60	Virgo	Elliptical galaxy	9.2	
61	Virgo	Spiral galaxy	9.6	
62	Ophiuchus	Globular	8.9	
63	Canes Venatici	Spiral galaxy	10.1	
64	Coma Berenices	Spiral galaxy	6.6	
65	Leo	Spiral galaxy	9.5	
66	Leo	Spiral galaxy	8.8	
67	Cancer	Open cluster	6.1	Famous old cluster
68	Hydra	Globular	9.0	
69	Sagittarius	Globular	8.9	

Number	Constellation	Type	Magnitude	Remarks
70	Sagittarius	Globular	9.6	
71	Sagitta	Globular	9.0	
72	Aquarius	Globular	9.8	
73	Aquarius	Four faint stars	–	Not a cluster
74	Pisces	Spiral galaxy	10.2	
75	Sagittarius	Globular	8.0	
76	Perseus	Planetary	12.2	
77	Cetus	Spiral galaxy	8.9	
78	Orion	Nebula	8.3	
79	Lepus	Globular	7.9	
80	Scorpio	Globular	7.7	
81	Ursa Major	Spiral galaxy	7.9	
82	Ursa Major	Irregular galaxy	8.8	
83	Hydra	Spiral galaxy	10.1	
84	Virgo	Spiral galaxy	9.3	
85	Coma Berenices	Spiral galaxy	9.3	
86	Virgo	Elliptical galaxy	9.7	
87	Virgo	Elliptical galaxy	9.2	Radio source
88	Coma Berenices	Spiral galaxy	10.2	
89	Virgo	Elliptical galaxy	9.5	
90	Virgo	Spiral galaxy	10.0	
92	Hercules	Globular	6.1	
93	Puppis	Open cluster	6.0	
94	Canes Venatici	Spiral galaxy	7.9	
95	Leo	Barred spiral galaxy	10.4	
96	Leo	Spiral galaxy	9.1	
97	Ursa Major	Planetary	12.6	Owl Nebula
98	Coma Berenices	Spiral galaxy	10.7	
99	Coma Berenices	Spiral galaxy	10.1	
100	Coma Berenices	Spiral galaxy	10.6	
101	Ursa Major	Spiral galaxy	9.6	
103	Cassiopeia	Open cluster	7.4	
104	Virgo	Spiral galaxy	8.4	"Sombrero Hat" Galaxy

Various forms of the Messier catalogue have been given, notably by Owen Gingerich (*Sky and Telescope*, Vol. XIII, p. 158 [1954]) and R.H. Garstang (BAA *Handbook*, p. 63 [1964]). Five additions were made, all objects observed by the French astronomer Méchain, and these are often included in the catalogue: M.105 (elliptical galaxy in Leo), M.106 (spiral galaxy in Canes Venatici), M.107 (globular in Ophiuchus), and M.108 and 109 (spiral galaxies in Ursa Major).

M.40 is not identifiable; it may simply be a couple of faint stars, or it may have been a comet. M.91 is also an absentee, and this too may have been a comet, though Gingerich suggests that it may be identical with M.58. There is grave doubt about the identities of M.47 and M.48; it has been suggested that M.47 is an open cluster in Argo Navis (Puppis) and M.48 an open cluster in Hydra. M.102 may have been identical with M.101, or it may possibly have been a faint spiral galaxy in Draco. Finally, M.73 consists of four faint, unconnected stars and is not a true cluster or nebular object.

The Caldwell Catalogue

Almost all the Caldwell objects are easy, but I have added also the number in the official New General Catalogue of Clusters and Nebulae. The list is arranged in declination, so that objects with numbers higher than about C. 70 will be too far south for observers in most of Europe.

C	NGC	Constellation	Type	Magnitude	Name
1	188	Cepheus	Open cluster	8.1	
2	40	Cepheus	Planetary nebula	10.7	
3	4236	Draco	Elliptical galaxy	9.6	
4	7023	Cepheus	Reflection nebula	–	
5	IC342	Camelopardalis	Spiral galaxy	9.2	
6	6543	Draco	Planetary nebula	8.1	Cat's eye Nebula
7	2403	Camelopardalis	Spiral galaxy	8.4	
8	559	Cassiopeia	Open cluster	9.5	
9	Sh2-155	Cepheus	Bright nebula	–	Cave Nebula
10	663	Cassiopeia	Open cluster	7.1	
11	7635	Cassiopeia	Bright nebula	8	Bubble Nebula
12	6948	Cepheus	Spiral galaxy	8.9	
13	457	Cassiopeia	Open cluster	6.4	Phi Cas. clust
14	869/884	Perseus	Double cluster	4.3	Sword-Handle
15	6826	Cygnus	Planetary nebula	8.8	Blinking Nebula
16	7243	Lacerta	Open cluster	6.4	
17	147	Cassiopeia	Elliptical galaxy	9.3	
18	185	Cassiopeia	Elliptical galaxy	9.2	
19	IC 5146	Cygnus	Bright nebula	–	Cocoon Nebula
20	7000	Cygnus	Bright nebula	–	N.America Nebula
21	4449	Canes Venatici	Irregular galaxy	9.4	
22	7662	Andromeda	Planetary nebula	8.3	Blue Snowball
23	891	Andromeda	Spiral galaxy	9.9	
24	1275	Perseus	Spiral galaxy	11.6	Perseus
25	2419	Lynx	Globular	10.4	

C	NGC	Constellation	Type	Magnitude
26	4244	Canes	Venatici	Spiral galaxy 10.2
27	6888	Cygnus	Bright nebula	– Crescent Nebula
28	752	Andromeda	Open cluster	5.7
29	5005	Canes	Venatici	Spiral galaxy 9.5
30	7331	Pegasus	Spiral galaxy	9.5
31	IC 405	Auriga	Bright nebula	– Flaming Star Nebula
32	4631	Canes Venatici	Spiral galaxy	9.3 Whale Galaxy
33	6992/5	Cygnus	Supernova remnant	– Eastern Veil Nebula
34	6960	Cygnus	Supernova remnant	– Western Veil Nebula
35	4889	Coma Berenices	Elliptical galaxy	11.4
36	4559	Coma Berenices	Spiral galaxy	9.8
37	6885	Vulpecula	Open cluster	5.9
38	4565	Coma	Berenices	Spiral galaxy 9,6 Needle Galaxy
39	2382	Gemini	Planetary nebula	8.6 Eskimo Nebula
40	3626	Leo	Spiral galaxy	10.9
41	Melotte 25	Taurus	Open cluster	0.5 Hyades
42	7006	Delphinus	Globular	10.6
43	7814	Pegasus	Spiral galaxy	10.3
44	7479	Pegasus	Spiral galaxy	10.9
45	5248	Boötes	Spiral galaxy	10.2
46	2261	Monocerus	Bright nebula	– Hubble's Variable Nebula
47	6934	Delphinus	Globular	8.9
48	2775	Cancer	Spiral galaxy	10.1
49	2237/9	Monoceros	Bright nebula	– Rosette Nebula
50	2244	Monoceros	Open cluster	4.8
51	IC 1613	Cetus	Irregular galaxy	9.2
52	4697	Virgo	Elliptical galaxy	9.3
53	3415	Sextans	Spiral galaxy	8.9 Spindle Galaxy
54	2506	Monoceros	Open cluster	7
55	7009	Aquarius	Planetary nebula	8.0 Saturn Nebula
56	246	Cetus	Planetary nebula	8.6
57	6822	Sagittarius	Irregular galaxy	8.8
58	2360	Canis Major	Open cluster	7.2
59	3242	Hydra	Planetary nebula	8.6 Ghost of Jupiter
60	4038	Corvus	Spiral galaxy	10.7)
61	4039	Corvus	Spiral galaxy	10.7) The Antennae
62	247	Cetus	Spiral galaxy	9.1
63	7293	Aquarius	Planetary nebula	6.3 Helix Nebula
64	2362	Canis Major	Open cluster	8 Tau Canis Majoris cluster
65	253	Sculptor	Spiral Galaxy	7.1
66	6594	Hydra	Globular	10.2
67	1097	Fornax	Spiral galaxy	9.2
68	6729	Corona Australis	Bright Nebula	– R CrA Nebula

C	NGC	Constellation	Type	Magnitude
69	6302	Scorpius	Planetary nebula	9.6
70	300	Sculptor	Spiral galaxy	8.1
71	2477	Puppis	Open cluster	5.8
72	55	Sculptor	Spiral galaxy	7.9
73	1854	Columba	Globular	7.3
74	3132	Vela	Planetary nebula	8.2
75	6124	Scorpius	Open cluster	5.8
76	6231	Scorpius	Open cluster	2.6
77	5128	Centaurus	Radio galaxy	6.8 Centaurus A
78	6541	Corona Australis	Globular	6.6
79	3201	Vela	Globular	6.7
80	5130	Centaurus	Globular	3.6 Omega Centauri
81	6352	Ara	Globular	7
82	6193	Ara	Open cluster	5.2
83	4945	Centaurus	Spiral galaxy	8.7
84	5286	Centaurus	Globular	7.6
85	IC 2391	Vela	Open cluster	2.5 Velorum cluster
86	6397	Ara	Globular	5.7
87	1261	Horologium	Globular	8.4
88	5823	Circinus	Open cluster	7.9
89	6087	Norma	Open cluster	5.4
90	2867	Carina	Open cluster	9.7
91	3532	Carina	Open cluster	3.0
92	3372	Carina	Bright nebula	– Eta Carinae Nebula
93	6752	Pavo	Globular	5.4
94	4755	Crux	Open cluster	4.2 Jewel Box cluster
95	6025	Triangulum Australe	Open cluster	5.3
96	2516	Carina	Open cluster	3.8
97	3766	Centaurus	Open cluster	5.3
98	4609	Crux	Open cluster	6.9
99		Crux	Dark Nebula	– Coal Sack
100	Coll.249	Centaurus	Cluster	4.5
101	6744	Pavo	Spiral galaxy	8.3
102	IC 2602	Carina	Open cluster	1.9 Theta Carinae cl
103	2070	Dorado	Bright Nebula	– Tarantula Nebula
104	362	Tucana	Globular	6.6
105	4833	Musca	Globular	7.3
106	104	Tucana	Globular	4.0 47 Tucanae
107	6101	Apus	Globular	6.3
108	4372	Musca	Globular	7.8
109	3195	Chamaeleon	Planetary nebula	8.4

The Star Maps

Many periodicals and some of the national newspapers give regular "stars of the month" charts. These are useful, but in my personal opinion they are of limited help to the absolute beginner, since they show so many objects that confusion is bound to result.

I have found that the best way to learn the various groups is to pick them out, one by one, by means of the two leading constellations of our skies, Ursa Major (the Great Bear) and Orion. Of these, Orion is the more brilliant, but it is not always visible in England, whereas the Bear never sets.

Using these two constellations as "signposts in the sky", it is possible to identify the other groups, and this system is developed in the maps given here. The key maps, Map I and II, will enable the beginner to find his way about in Maps IV to X. There can be little difficulty in finding Orion and the Great Bear; for one thing, there will always be someone nearby who knows them.

The star maps given here are not precision charts; nor are they intended to be, but it is hoped that they will be of some use as an aid to finding one's way about the sky.

In the constellation notes, all stars down to magnitude 3.5 have been listed under the heading "Chief Stars". All the doubles, variables and clusters mentioned are easy objects.

The Northern Stars

Map I. Key Map: Ursa Major (The Great Bear)

Almost everyone must know the Great Bear. Its seven stars are a familiar feature of the night sky, and it is of course so far north that it never sets in the latitude of England. The proper names of the seven are frequently used: in addition, Merak and Dubhe are popularly known as the "Pointers".

The first step after having identified the Bear is to find the Pole Star. Imagine a line drawn from Merak through Dubhe, and extended; it will reach a second-magnitude star rather "out on its own", and this is Polaris. The Little Bear, Ursa Minor, can then be picked out, bending back towards the Great Bear itself. The stars are much fainter, but one of them, the rather reddish Kocab, is of magnitude 2.

Now imagine a line from Alioth, in the Great Bear, through Polaris. Extended for an equal distance on the far side of Polaris, it will reach five brightish stars (magnitudes 2 to 3) arranged in a rough W. This is Cassiopeia, which, like the Bears, never sets in England.

A line from Megrez through Dubhe will come eventually to Capella, which is one of the brightest stars in the entire sky. It is circumpolar in England, but at it lowest, as during summer evenings, it almost reaches the horizon. In winter evenings it is high up and may indeed pass overhead. If you see a really bright star straight above you, it can only be Capella or Vega; Capella is yellowish and may be recognized by the small triangle of stars close by it, whereas Vega is decidedly blue. Vega can be found by means of a line beginning at Phad, passing between Megrez and Alioth, and extended for some distances across the sky.

The remaining stars shown in Map I are not circumpolar. The Twins, Castor and Pollux, may be found by means of a line from Megrez through Merak; they are at least at their best in winter. Regulus and the other stars of the Lion, found by a line from Megrez through Phad, seem to follow the Twins in the sky; the curved arrangement of stars rather like a reversed question mark, of which Regulus is the brightest, is known as the "Sickle of Leo", and is easy to recognize. Even easier is Arcturus, about as bright as Capella and Vega. This is found by means of a line from Mizar through Alkaid, and curved rather downwards; if the curve is continued through Arcturus it comes to another first-magnitude star, Spica, in Virgo.

Arcturus and Spica are prominent features of the spring and summer skies of England.

It may be added that Arcturus shines with a distinctly orange light, so that it cannot be confused with Capella or Vega.

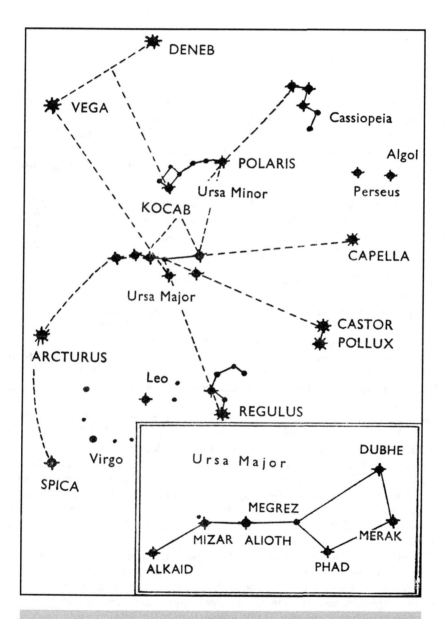

MAP I

Map II. Key Map: Orion

It is a pity that Orion is not circumpolar in England, as it is a magnificent "signpost", as well as being a beautiful constellation in itself. It cannot be mistaken, as all its chief stars are brilliant, two of the first magnitude (Betelgeux and Rigel) and five of the second magnitude. Mintaka, Alnilam and Alnitak form the famous Belt. The periods of visibility of Orion in England can be judged from the following:

January 1st	Rises 4 p.m., highest 10 p.m., sets 5 a.m.
April 1st	Rises in daylight, highest in daylight, sets 11 p.m.
July 1st	Rises 4 a.m., highest in daylight, sets in daylight.
October 1st	Rises 10 p.m., highest 5 a.m., sets in daylight.

It must be understood that these times are only very rough; Orion covers a considerable area, and takes some time to "rise". It is, however, clear that the constellation is best seen in winter and in early mornings in autumn.

The first-magnitude stars in the key map are easy to find if Orion can be seen. The three stars of the Belt (Mintaka, Alnilam, and Alnitak) point downwards to Sirius, which is the most brilliant star in the sky, though of course less bright than Venus, Jupiter, or Mars when well placed. Upwards, the Belt stars indicate Aldebaran in Taurus, a reddish first-magnitude star of about the same colour and brightness as Betelgeux.

Bellatrix and Betelgeux point more or less to Procyon, in Canis Major, which is not much fainter than Rigel; if this line is continued and curved slightly, it reaches a reddish second-magnitude star, Alphard, in Hydra, known as "the Solitary One" because it lies in a very barren region. The Twins, Castor and Pollux, can be found by a line from Rigel through Betelgeux; because they can also be found by using Ursa Major, this links the two key maps. Capella is indicated by a line from Saiph through Alnitak. Diphda in Cetus, the other star shown in the diagram, is less easy to find. It is only of magnitude 2, and is frequently visible when Orion is below the horizon.

Undoubtedly, a winter evening is the best time to start star recognition, because then both our "signposts", Orion and the Bear, can be seen. If a start is to be made in summer, we must do without Orion; but the Bear can by itself teach us the way about the heavens, and even though the stars seem at first to be arranged in a chaotic manner, it takes surprisingly little time to find one's way about.

Each of the following charts contains at least one key map object. Exact positions of telescopic objects, in right ascension and declination, are not given here, because an observer who possesses a telescope equipped with setting circles will in any case need a more detailed set of charts.

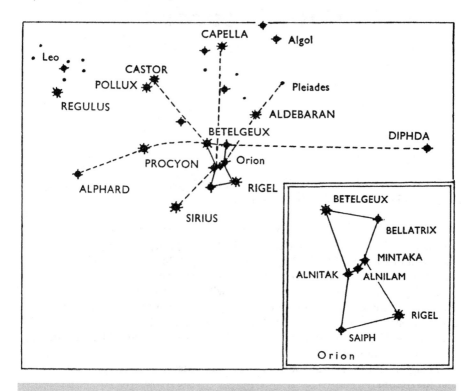

MAP II

Map III. Ursa Major, Ursa Minor, Draco, Cepheus, Camelopardus

This is the North Polar region. The stars in it are of course circumpolar in England and will quickly be recognized.

URSA MAJOR. This has already been described at length. The chief stars are Epsilon (Alioth) and Alpha (Dubhe) (1.8), Eta (Alkaid) (1.9), Zeta (Mizar) (2.1), Beta (Merak) and Gamma (Phad) (2.4), Psi and Mu (3.0), Iota (3.1), Theta (3.2), Delta (Megrez) (3.3), and Lamda (3.4). Part of the constellation extends onto Map VI.

Double Star. Zeta (Mizar). Naked-eye pair with Alcor. In low power, Mizar is itself double: magnitudes 2.2, 3.9; distance 14".5; P.A. 150°. Between this pair and Alcor is another star. Nu; magnitudes 3.7, 9.7; distance 7".2: PA 147°.

Variables. T: magnitude 5.5 to 13, period 257 days. Spectrum Me: red. An easy object near maximum.

R: magnitude to 13; period 302 days. Spectrum Me: red. K. Like T, an easy object near maximum.

Clusters and Nebulae. M.81 and M.82; two galaxies, close together, identifiable without much difficulty.

M.97: The Owl Nebula, a planetary so called because its two hot stars do give it the look of an owl's face with high powers. It is very faint with small apertures but is worth looking for.

URSA MINOR curves down over the stars of Ursa Major. Chief stars: Alpha (Polaris) (2.0), Beta (Kocab) (2.0), Beta (Kocab) (2.0), and Gamma (3.1). Kocab is a fine orange star.

Double Star. Polaris: Magnitudes 2.0, 9.0; distance 18".3, P.A. 217°. An easy object with aperture 3 inch or more.

DRACO. A long, winding constellation, stretching from Lambda (between Dubhe and Polaris) as far as Gamma, which lies near Vega. The chief stars are Gamma (2.2), Eta (2.7), Beta (2.8), Delta (3.1), Zeta (3.2) and Iota (3.3). Alpha or Thuban (3.6) used to be the pole star in ancient times.

Double Stars. Nur: magnitudes 4.5, 4.5; distance 62". This is a very wide, easy double.

Eta: magnitudes 2.7, 8.0; distance 6″; P.A. 142°. This can be seen with a 3-inch refractor.

Epsilon: magnitudes 4.0, 7.5; distance 3″.3; P.A. 009°.

CEPHEUS is not one of the easier constellations to identify, but it is useful to remember that Gamma Cephei lies more or less between Polaris and the "W" of Cassiopeia. It is better shown on Map VII. Chief stars: Alpha (2.4), Beta (3.1), and Gamma (3.2). Telescopic objects are given in the notes on Map VII.

CAMELOPARDUS. A large, dull constellation, with no stars brighter than the fourth magnitude, and with no objects of particular interest. It is in fact one of the most barren regions of the heavens.

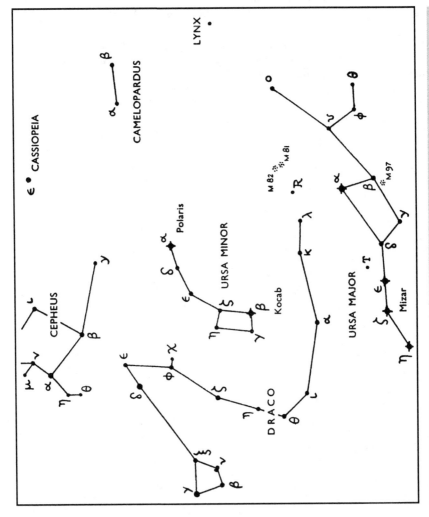

MAP III

Map IV. Orion, Lepus, Eridanus, Taurus, Cetus, Auriga, Columba, Caelum, Fornax

The times of rising and setting of Orion, in England, were given in the notes on Map II. Capella is just circumpolar but can almost graze the horizon. Perseus is shown in part and also Triangulum.

ORION is probably the most glorious constellation in the heavens, and is easy to recognize. Betelgeux is a fine sight with a lower power (spectrum M; orange-red), and Rigel is brilliantly white. Rigel appears only very slightly less brilliant than Arcturus and Vega. The other leading stars are Gamma (Bellatrix) (1.6), Epsilon (Alnilam) (1.7), Zeta (Alnitak) (1.8), Kappa (Saiph) (2.1), Delta (Mintaka) (2.3, but slightly variable), Iota (2.8) Pi3 (3.2), Eta (3.4), and Lambda (3.5).

Double Stars. Rigel: magnitudes 0.1, 7.0; distance 9".2; P.A. 206°. A test for a 2-inch O.G.; easy with a 3-inch. The companion is said to be bluish, but to me it always appears white.

Eta; magnitudes 3.6, 4.8; distance 1".5; P.A. 083°

Lambda: magnitudes 3.6, 5.5; distance 4".4; P.A. 043°.

Zeta: magnitudes 1.9, 4.2; distance 2".4; P.A. 164°. I find this very hard with anything less than 3-inch aperture.

Iota: magnitudes 3.0, 7.4; distance 11".4; P.A. 142°. Immersed in nebulosity.

Theta: the Trapezium, a multiple star. Magnitudes 6.0, 7.0, 7.0, 7.5. All four stars are easy in a 3-inch O.G. Immersed in the Great Nebula, M.42.

Sigma: another multiple. The magnitudes of the four brightest stars are 4.0, 7.0, 7.5, and 9.9. Less striking than Theta, but well worth examination.

Delta: magnitudes 2.3 (var.) 6.7; distance 53"; P.A. 000°. Very wide and easy.

Variables. Betelgeux 0.0 to 1.0. This is a greater range than is given in most textbooks, but Sir John Herschel recorded that he saw it outshine Rigel, and this has also been my experience. The best comparison star for normal periods is of course Aldebaran; another, useful when Betelgeux is faint, is Pollux.

U (not far from Zeta Tauri): magnitudes 5.5 to 12.6. Period 372 days. A red, Me-type long-period variable.

Delta (Mintaka): an eclipsing binary of small magnitude range (2.20 to 2.35).

Clusters and Nebulae. M.42: the Sword of Orion, visible to the naked eye, and the most prominent of all galactic nebulae. It is a splendid sight in a small telescope; dark nebulosity may be seen close to the Trapezium.

LEPUS is a small constellation near Orion. The chief stars are Alpha (2.6), Beta (2.8), Epsilon (3.2), and Mu (3.3).

Double Stars. Kappa: magnitudes 4.9, 7.5; distance 2".6; P.A. 000°. The primary is yellowish and the companion bluish.

Beta: magnitudes 2.8, 9.4; distance 2".5; P.A. 313°.

Variable. R: magnitude 5.9 to 10.5; period 432 days. This is an intensely red star of spectrum N. It is not hard to find when near maximum.

COLUMBA lies below Lepus and is too far south to be well seen in England. Chief stars: Alpha (2.6), Beta (3.1). The constellation contains no features of particular interest.

CAELUM has no star brighter than Alpha (4.5) and is always very low in our latitudes.

Double Star. Gamma: magnitudes 4.7, 8.5; distance 3"; P.A. 310°.

ERIDANUS. A very long constellation, of which the chief stars are the first-magnitude Achernar, and Beta (2.8), Theta (2,9), and Gamma (3.0). Achernar and Theta never rise in England but are shown in Map XV.

Double Stars. Omicron^2: magnitudes 4.0, 9.0; distance 82": P.A. 107°. Theta is double; separation 8".

FORNAX has no star brighter than Alpha (4.0). It is low in England and contains no features of interest.

CETUS is another long, winding constellation; the rest of it is shown in Map X. Chief stars: Beta (2.0), Alpha (2.5), and Eta, and Tau (3.5). Alpha is a fine orange star.

Double Stars. Gamma: magnitudes 3.6, 6.2; distance 3"; P.A. 295°.

66: magnitudes 6.0, 7.8; distance 16".3; P.A. 232°. The primary is yellow and the companion blue. This is in a low-power field with Mira, and is a useful guide when Mira is faint.

Variable. Omicron (Mira): magnitude 1.7 to 9.6; period 331 days. This interesting star is fully described in the text.

Nebula. M.77: a fairly easy object, one degree away from Delta. It is actually a spiral galaxy.

TAURUS. This is a Zodiacal constellation of great interest. Apart from Alderbaran, the chief stars are Beta (1.6), Eta (Alcyone) (2.9), Zeta (3.1), and the two of the Hyads, Theta2 (3.4) and Epsilon (3.5). Beta Tauri was once known as Gamma Aurigae.

Double Star. Aldebaran has a 13th magnitude companion; distance 121"; P.A. 034°. This is a wide optical double, but the faintness of the companion makes it a useful test object.

Variable. Lambda: magnitude 3.3 to 4.2; period 3.9 days. Spectrum B3. this is an eclipsing binary of the Algol type.

Clusters and Nebulae. M.1: the remarkable "Crab Nebula", near Zeta, described in the text.

The Pleiades and Hyades are also described in the text. The Hyades, which are scattered, are best seen in binoculars.

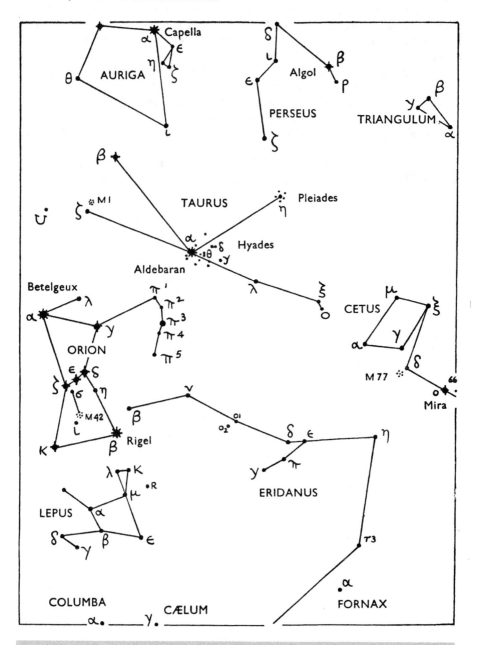

MAP IV

AURIGA. On of the brightest of the northern groups. Capella is shown in both key maps and is surpassed by only three other stars visible from England: Sirius, Vega, and Arcturus. The difference between Vega and Capella is only 1/100 of a magnitude. Capella is yellow, and can be identified by the three fainter stars (Epsilon, Zeta, Eta) close by it; these have been termed the "Haedi", or Kids. Gamma Aurigae is now known as Beta Tauri. This is one of a few cases of stars being included in two constellations; others are Alpha Andromedae (= Delta Pegasi) and Gamma Scorpii (= Sigma Librae).

The other chief stars of Auriga are Beta (1.9), Iota, and Theta (2.6) and Eta (3.2). Epsilon, the vast giant, is variable; it is comparable with Eta. The magnitude range is small, and this applies also to the other giant eclipsing binary, Zeta, whose fluctuations will not easily be detected without instruments.

Epsilon's magnitude varies from 3.1 to about 4.4, and the period is just over 27 years.

Double Star. Theta: magnitudes 2.6, 7.1; distance 2".8; P.A. 333°. I always find this rather difficult with a 6-inch reflector; it is said to be a test for a 4-inch O.G.

Map V. Gemini, Cancer, Canis Major, Canis Minor, Monoceros, Hydra

The constellations shown in this map are at their best in winter and spring evenings. The following times of rising and setting in England are for Cancer, and are of course very rough. Cancer is a Zodiacal constellation, as are Gemini and Leo.

January 1st	Rises 6 p.m., highest 2 a.m., sets in daylight.
April 1st	Rises in daylight, highest 8 p.m., sets 4 a.m.
July 1st	Rises in daylight, highest in daylight, sets in daylight.
October 1st	Rises at midnight, highest 8 a.m., sets in daylight.

CANIS MAJOR is most notable because of the presence of Sirius, the brightest star in the sky. The other chief stars are Epsilon (1.5), Delta (1.8), Beta (2.0), and Omicron^2 and Zeta (3.0). The group is easy to find from Orion. Actually there are few interesting telescopic objects in Canis Major, but M.41 is a bright cluster well worth looking at.

CANIS MINOR contains Procyon; the only bright star is Beta (2.9).

MONOCEROS. A large, faint constellation with no star brighter than the fourth magnitude; it lies in the area enclosed by Procyon, Sirius, and Betelgeux. The Milky Way passes through it, and there are some rich telescopic fields, so that the region is worth sweeping with low powers.
Double Star. Beta: a triple. Magnitudes 5.0, 5.5, 5.9; distances 7″.4 and 2″.8; P.A.s 132° and 105°.
Cluster. Around the sixth-magnitude star 12 Monocerotis is a fine open cluster, H.VII.2 (not in Messier's catalogue). It lies between Betelgeux and the fourth-magnitude star Delta Monocerotis.
A few stars of PUPPIS, including Rho (2.7) and Xi (3.3), can be made out, and a few stars of PYXIS NAUTICA can also be seen low down on the horizon.

Cluster. M.46: a beautiful small cluster, roughly between Rho Puppis and Alpha Monocerotis.

GEMINI. This is one of the grandest of all constellations. As well as Pollux and Castor, it includes other bright stars; Gamma (1.9), Mu and Epsilon (3.0), Xi (3.4), and Delta (3.5). Moreover, the Milky Way passes through it. Castor and Pollux can be found from either key map. Pollux, rather orange in colour (type K), is now appreciably brighter than Castor, though it may be that in Ptolemy's day this was not the case. There are plenty of interesting objects in Gemini.

Double Stars. Castor: magnitudes 1.9, 2.8; distance 1.8″; P.A. 151°. A fine double; binary, period 380 years. As described in the text, Castor is a multiple system; Castor C, magnitude 9.1 lies at 73″.

Delta: magnitudes 3.5, 8.2; distance 6″.7; P.A. 210°. Test for a 2-inch O.G., though I always find it rather easy with such an aperture.

Lambda: magnitudes 3.7, 10.0; distance 10″; P.A. 033°.

Kappa: magnitudes 4.0, 8.5; distance 6″.7; P.A. 235°.

Variables. Eta: a long-period M-type variable: magnitude 3.3 to 4.2; official period 231 days, but I have my doubts!

Zeta: magnitude 3.7 to 4.3; period 10.2 days. Spectrum G. A typical Cepheid. A useful comparison star is Nu (4.1).

R: magnitude 6.0 to 14; period 370 days.

Cluster. M.35. A fine open cluster, a splendid sight in a small telescope. Mu and Eta act as an excellent "guide stars" to it.

CANCER. A faint constellation, the brightest stars being Beta (3.8), Iota (4.1), and Delta (4.2), but it incudes some interesting objects, such as Praesepe. It is not unlike a very dim and ghostly Orion and lies in the area enclosed by Pollux, Procyon and Regulus.

Double Stars. Zeta: magnitudes 5.0, 5.7; distance 1″; binary, period 60 years. It was at its widest in the year 1960 and is now closing up again. There is a third component: magnitude 6.1, distance 5″.7.

Iota: magnitudes 4.3, 6.3; distance 31″; P.A. 307°. The larger star is yellowish, the companion bluish.

Variable. R Magnitude 5.9 to 11.5; period 362 days. An M-type, long-period variable.

Clusters. M.44 (Praesepe). One of the best of the open clusters. It has been described in the text and can be seen with the naked eye on any reasonable transparent moonless night.

M.67. A conspicuous telescopic object close to Alpha (4.3).

HYDRA. Apart from Argo Navis, which is now divided up, Hydra is the largest constellation in the sky; parts of it are also shown on Maps VI and IX. It is however rather barren. The chief stars are Alpha (2.0), Gamma (3.0), Zeta and Nu (3.1), Pi (3.2), and Epsilon (3.4). Alphard is shown on the Key Map II; it is easy to find, as it is distinctly reddish and appears very isolated. Its name of "the Solitary One" suits it well. It can be identified by continuing the "sweep" from Bellatrix through Betelgeux and Procyon and, incidentally, Castor and Pollux point to it. It is a fine object in a low power.

Double Star. Theta: magnitudes 4.9, 10.8; distance 38″; P.A. 185°. The faintness of the companion makes it a useful test.

Epsilon: magnitudes 3.6, 7.7; distance 3″.6; P.A. 253°. This is in the "head" of Hydra, which is easy to find, as it lies roughly midway between Procyon and Regulus. The bright component is a close binary with a period of 15 years. A magnitude 12 star lies at 20″.

SEXTANS. A faint and unremarkable constellation, with no star as bright as the fourth magnitude and no interesting telescopic objects.

Parts of Leo and Taurus are included in this map but are better shown in Maps VI and IV, respectively.

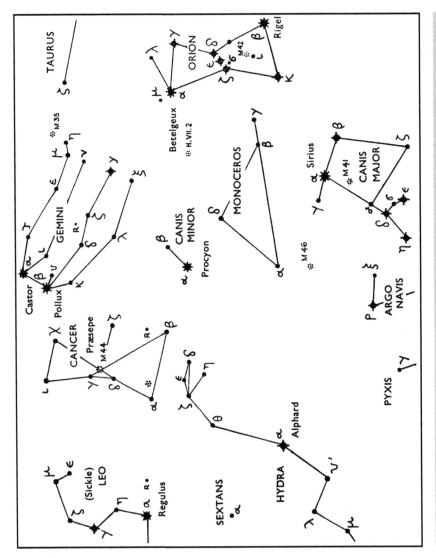

MAP V

Map VI. Leo, Virgo, Coma Berenices, Corvus, Crater, Leo Minor

This area contains some interesting features; Regulus, Spica, and Arcturus are shown in Map Key I. The rough times of rising and setting for Spica, in England, are:

January 1st	Rises 1 a.m., highest 6 a.m., sets in daylight.
April 1st	Rises 7 p.m.. highest midnight, sets 5 a.m.
July 1st	Rises in daylight, highest in daylight, sets 11 p.m.
October 1st	Rises in daylight, highest in daylight, sets in daylight.

LEO. A large, important constellation. Regulus is of course the chief star, and other bright stars are Gamma (2.0), Beta (2.1), Delta (2.6), Epsilon (3.0), Theta (3.3), and Zeta (3.5). The curved line of stars beginning with Regulus is known as the Sickle, and is a prominent feature; the triangle formed by Beta, Delta, and Theta is also easy to find. Beta is a secular variable. Ptolemy made it of the first magnitude, but it is now below the second. As it is suspected of variability, it is well worth watching; Gamma makes a good companion star.

Double Stars. Gamma: magnitudes 2.3, 3.8; distance 4″.3; P.A. 121°. A fine binary, with a period of 407 years.

Iota: magnitudes 3.9, 7.0; distance 0″.6; P.A. 015°.

Variable Star. R: magnitude 5.0 to 10.5, period 312 days. A long-period M-type variable, visible to the naked eye near maximum.

LEO MINOR. A faint group, about midway between Regulus in Leo and Merak in Ursa Major. It contains no star brighter than magnitude 4. The only object of interest is the M-type long-period variable R; magnitude 6.2 to 12.3, period 370 days.

VIRGO. In shape, Virgo is rather like a roughly drawn Y. The brightest star is of course Spica; others are Gamma (2.8), Epsilon (2.9), and Zeta (3.4). The "bowl" of the Y, in the area enclosed by it and Beta Leonis, is very rich in faint galaxies and is well worth sweeping.

Double Stars. Gamma: magnitudes 3.6, 3.7; distance about 5″. This is a magnificent binary, with a period of 172 years, and one of the best of all double stars for small telescopes.

Theta: magnitudes 4.0, 9.0; distance 7″; P.A. 343°. There is a 10th magnitude star at a distance of 71″, making a useful test.

Variables. R: magnitude 5.9 to 12.0; period 145 days.

S: magnitude 5.6 to 12.3; period 372 days. Like R, a long-period variable of spectrum M.

COMA BERENICES and CANES VENATICI lie in the area enclosed by the Great Bear, Regulus, Beta Leonis, and Arcturus. Coma contains no star brighter than magnitude 4½, but it is a rich area, and to the naked eye looks almost like a very scattered star-cluster, so that it is worth sweeping. Canes Venatici has one star, Alpha, of magnitude 2.9; it is a wide optical double: magnitudes 3.0, 5.6; distance 20″, P.A. 228°. Canes Venatici is so far north that it is circumpolar in England.

There are many clusters and nebulae in these two constellations, which are shown on the map and are well worth looking for.

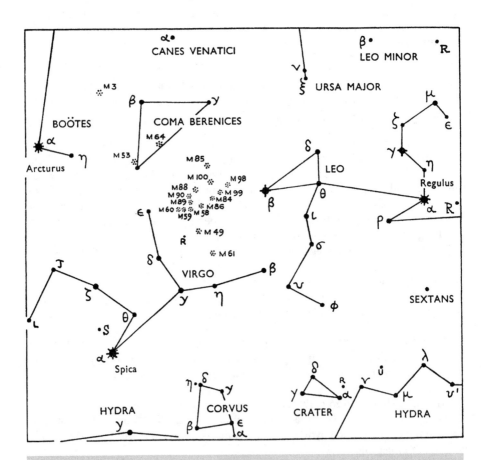

MAP VI

BOÖTES is shown in part, but is described with Map IX.

HYDRA is also partly shown, the brightest star being Nu (3.1). In this part of the constellation lies the interesting red N-type irregular variable U Hydrae, which has a magnitude range from 4.5 to 5.9.

CRATER. The brightest stars in this small group are Delta (3.8), Gamma (4.1), and Alpha (4.2), which form a triangle not far from Nu Hydrae. Not far from the reddish Alpha is the very red irregular variable R Crateris, with a magnitude range of from 8 to 9.

CORVUS. This is easy to find, as its four chief stars are of about the third magnitude (Gamma, 2.6; Beta, 2.7; Delta and Epsilon, each 3.0) and form a quadrilateral. To find it, pass a line from Arcturus midway between Spica and Gamma Virginis, the double star at the branch of the "Y". Delta Corvi is a double: magnitudes 3.1, 8.2; distance 24″; P.A. 212°.

Map VII. Cassiopeia, Cepheus, Lacerta, Perseus, Andromeda, Lynx, Triangulum

Of the groups in this map, all are circumpolar apart from sections of Andrdomeda and Triangulum. Ursa Minor and Camelopardus are also shown but are described with Map III.

CASSIOPEIA, shown in Key Map I, is one of the most interesting and conspicuous of the northern constellations. The Milky Way passes through it, and there are many rich telescopic fields. Of the chief stars, Alpha (Shedir) and Gamma are variable; the others are Beta (2.3), Delta (2.7), and Epsilon (3.4), which of course serve as excellent comparison stars.

Double Stars. Alpha: magnitudes 2.2 (var.) and 9.0; distance 63″; P.A. 280°. A wide optical double.

Eta: magnitudes 3.7, 7.3; distance 11″.2; P.A. 298°. Binary.

Iota: a fine triple. Magnitudes 4.2, 7.1, 8.0; distances 2″.3, 7″.5; P.A.s 251°, 113°°.

Variables. Gamma: magnitude 1.7 to 3.4: irregular, and now classed as a "pseudo-nova". A most peculiar star, with a most unusual spectrum. Between 1965 and 2005, its magnitude averaged around 2.3. It is well worth watching.

Alpha. This was long classed as a variable. Recently, doubts have been cast on the reality of the fluctuations, but my own rough observations between 1936 and the present time indicate that the magnitude fluctuates irregularly between 2.1 and 2.5. Also worth watching is the irregular Rho.

R: magnitude 5.3 to 13.0; period 432 days. Spectrum M.

Clusters and Nebulae. M.52: a fairly bright cluster. Alpha and Beta act as "guides" to it.

M.103; An open cluster close to Delta.

CEPHEUS. This is not too easy to identify. The chief stars are Alpha (2.4), Beta (3.1), Gamma (3.2), Zeta (3.3), and Eta (3.4). Gamma lies between Beta Cassiopeiae and Polaris; the main part of the constellation between Cassiopeia and Vega. The triangle made up of Zeta, Delta, and Epsilon is the most conspicuous feature. On the whole, Cepheus is rather a barren group.

Double Stars. Beta: magnitudes 3.3, 8.0; distance 14″; P.A. 250°.

Kappa: magnitudes 4.0, 8.0; distance 7".5; P.A. 122°.

Variables. Delta: Magnitude 3.5 to 4.4; period 5.37 days. The prototype Cepheid.

Mu: magnitude 3.6 to 5.1; irregular. Sir William Herschel's "garnet star". It is of type M and is probably the reddest of the naked-eye stars; a splendid object in a low power.

T: magnitude 5.5 to 9.6; period 391 days. Spectrum M.

AR; magnitude 7.1 to 7.8; period 116 days. Semi-regular.

LACERTA is a small constellation near Cepheus. It contains no star brighter than the fourth magnitude, and no objects of special interest.

PERSEUS. A grand constellation. It lies between Cassiopeia and Aldebaran; the chief star, Alpha (1.8) can be found by a line drawn from Gamma Cassiopeiae through Delta Cassiopeiae and extended. The other leading stars are Beta (Algol) (variable; 2.1 at maximum), Zeta (2.8), Epsilon and Gamma (2.9), Delta (3.0), and Rho (variable; 3.2 at maximum). The Milky Way is particularly rich in Perseus.

Double Stars. Zeta: magnitudes 2.8, 9.4; distance 12".5; P.A. 208°. The chief component is a very luminous B1-type super-giant.

Eta: magnitudes 4.0, 8.5; distance 28".4: P.A. 300°. The primary is yellow, the companion bluish.

Epsilon: magnitudes 2.9, 8.3; distance 9"; P.A. 009°.

Variable Stars. Beta (Algol); magnitude 2.1 to 3.3. The prototype eclipsing binary, fully described in the text.

Rho: magnitude 3.2 to 4.2; an M4-type irregular. A suitable comparison star is Kappa, magnitude 4.00.

Clusters M.34; a fine open cluster, roughly between Kappa Persei and Gamma Andromedae, visible to the naked eye on a transparent night.

H.VI.33 and 34. The "Sword-Handle" clusters, described in the text. They are visible to the naked eye, and in my view are the most beautiful of all open clusters. Between them is a faint red star.

ANDROMEDA. This is a bright constellation, the leading stars being Beta (2.0), Alpha and Gamma (2.1), and Delta (3.2). Alpha is included in the Square of Pegasus and is also known as Delta Pegasi. It can be found by means of a line drawn from Epsilon Cassiopeiae through Delta Cassiopeiae and extended.

Double Stars. Gamma: magnitudes 2.5, 5.0; distance 9".8; P.A. 060°. A grand double, the components being yellow and blue. The small star is again double; magnitudes 5.4, 6.2; distance 0".7; P.A. 109°.

Variable. R: magnitude 5.9 to 15; period 410 days. A long-period M-type variable, too faint at minimum for small apertures. It lies near Theta Andromedae (4.4).

Galaxy. M.31; the Great Spiral, described in the text. It is visible to the naked eye as a misty patch close to Nu Andromedae (4.4), but a telescope of large size is needed to show its structure.

TRIANGULUM. A fairly conspicuous little group near Andromeda, the leading stars being Beta (3.0) and Alpha (3.4).

Variable. R: magnitude 5.8 to 12; period 266 days. Spectrum M.

Galaxy. M.33. A large but rather faint and ill-defined object, roughly between Alpha Trianguli and Beta Andromedae.

LYNX. One of the most barren of all constellations. It adjoins Camelopardus, and lies between Ursa Major and the Twins (Castor and Pollux). There are no bright stars or interesting telescopic objects worthy of mention here.

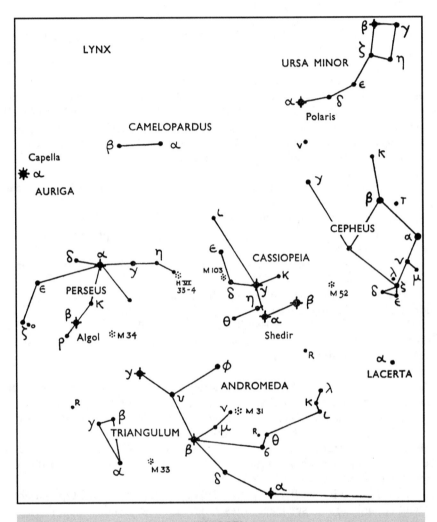

MAP VII

Map VIII. Cygnus, Lyra, Sagitta, Vulpecula, Delphinus, Equuleus, Capricornus, Aquila, Sagittarius, Scutum, Serpens, Aquarius

This is a very rich area, best seen in summer. Vega and Deneb are just circumpolar in England, and the approximate time of rising and setting for Altair are given below. It must be remembered that in all these "rising and setting" tables, allowance must be made for Summer Time.

January 1st   Rises 6 a.m., highest in daylight, sets 8 p.m.

April 1st   Rises midnight, highest 7 a.m., sets in daylight.

July 1st   Rises daylight, highest 1 a.m., sets in daylight.

October 1st   Rises in daylight, highest 7 p.m., sets 2 a.m.

Vega and Deneb are shown on the first key map. Vega is almost overhead at midnight near midsummer and can be recognized by its brilliance and by its bluish colour, which differs strongly from the yellowish hue of Capella, which occupies the overhead position at times during the winter.

LYRA. Though Lyra is a small constellation, and Vega is the only star above the third magnitude, it is remarkably rich in telescopic and other interesting objects. After Vega, the leading stars are Gamma (3.2) and the eclipsing Beta. The quadrilateral made up of Beta, Gamma, Delta, and Zeta is easily recognized.

Double Stars. Epsilon: the famous double-double, described in the text. The two main components can be split with the naked eye; magnitudes 4.5 and 4.7; distance 208″. Epsilon[1]; magnitudes 4.6, 6.3; distance 2″.8; P.A. 001°. Epsilon[2]: magnitudes 4.9, 5.2; distance 2″.2; P.A. 099°. Of the two, Epsilon[1] is the easier to divide, but both pairs are well visible in a 3-inch O.G.

Zeta: magnitudes 4.3, 5.9; distance 44″; P.A. 150°. A wide, easy double.

Eta: magnitudes 4.5, 8.0; distance 28″; P.A. 083°.

Vega: has a companion, magnitude 10.5, at a distance of 56″ and a P.A. of 169°. This is an optical pair, not a binary system. The faintness of the companion makes it a convenient test object.

Variables. Beta: magnitude 3.4 to 4.4, period 12.9 days. Eclipsing binary, described in the text. Gamma is a good comparison star; others are Zeta (4.1) and Kappa (4.3). It may be added here that the magnitude of 4.1 for Zeta as seen with the naked eye is the result of the combined light of the 4.3 and 5.9 magnitude components.

R: magnitude 4.0 to 5.0; a red M-type semi-regular variable.

Nebulae. M.57. Planetary. The Ring Nebula, described in the text. It can be seen with a small aperture, but the central star is extremely difficult even with large instruments. The object is easy to find, as it lies directly between Beta and Gamma Lyrae.

CYGNUS. The Swan, but also, and perhaps more appropriately, known as the Northern Cross. It is a superb constellation, in a rich part of the Milky Way. The chief star is Deneb; other bright stars are Gamma (2.2), Epsilon (2.5), Delta (2.9), Beta (3.1), and Zeta (3.2). It is worth remembering that Beta, the faintest of the stars forming the Cross, lies roughly between Vega and Altair.

Double Stars. Beta (Albireo): magnitudes 3.1, 5.1; distance 34″.6; P.A. 055°. Yellow primary, blue companion. I regard this as the loveliest double in the sky, and it is a superb object in any small telescope.

Delta: magnitudes 3.0, 6.5; distance 2″; P.A. 240°. A well-known test. Binary, with a period of 321 years.

61: magnitudes 5.6, 6.3; distance 28″; P.A. 142°. The celebrated star that was the first to have its distance measured.

Zeta: magnitudes 3.3, 7.9; distance 2″.3. Binary; period 500 years.

Variables. Chi: magnitude 4 to 14; period 409 days. A good comparison star when Chi is near maximum is its companion Eta (4.03).

W: magnitude 5.0 to 7.0; irregular. An M-type variable. It lies close to Rho (4.2).

X: magnitude 6.0 to 7.0; period 16.4 days. A Cepheid, lying close to Lambda (4.5).

The famous variables U, R, and SS Cygni are described in Appendix 26.

Nebulae and Clusters. There are many nebular objects in Cygnus. One of the most striking is M.39, near Rho, a good open cluster and a fine sight in a small telescope.

VULPECULA is a small constellation near Cygnus. It contains no star brighter than magnitude 4½. The most interesting object is M.27, the Dumbbell Nebula, a planetary; it is dim, but is well worth looking at, even though a telescope of some size is needed to show it properly. It lies not far from Gamma Sagittae. Vulpecula, the Fox, was once known as Vulpecula et Anser, the Fox and Goose; but nowadays the goose seems to have been discarded – possibly the fox has eaten it!

DELPHINUS. A beautifully compact little group, very easy to recognize. The brightest star is Beta (3.7). The most interesting object is the double star Gamma; magnitudes 4.5, 5.5; distance 10″.5; P.A. 270°. The primary is yellow, the companion green. The variables U and EU are described on page 267.

SAGITTA. Another compact group; the brightest stars are Gamma (3.7) and Delta (3.8). It lies between Altair and Beta Cygni.

EQUULEUS. The chief star of this little constellation is Alpha (4.1). Delta is an excessively close double, and a rapid binary.

PEGASUS. Most of this constellation, including the Square, is shown on Map X. The chief star in the present map is Epsilon (2.3), which is suspected of variability. Close to it lies the bright globular cluster M.15, a fine sight in a moderate telescope.

AQUARIUS also lies mainly in Map X. On the present map are Beta (2.9), Alpha (3.0), and two nebulus objects; the fine globular M.2, which I find fully resolvable with my 12½-inch. reflector and which lies between Beta Aquarii and Epsilon Pegasi, and the beautiful planetary NGC 7293 which lies in the same low-power field as the orange star Nu Aquarii (4.5). Aquarius is a Zodiacal constellation.

AQUILA. the chief stars, Altair, is one of the first magnitude and is easy to recognize because it has a brightish star to either side, Beta and Gamma. As well as Altair, the constellation includes Gamma (2.7), Zeta (3.0), Theta (3.1), Delta and Lambda (3.4) and Beta (3.9). The line below Altair, made up of Theta, Eta, and Delta, is very easy to identify.
Variables. Eta: magnitude 3.7 to 4.5, period 7.2 days. A typical Cepheid.
R: magnitudes 5.7 to 12; period 300 days. Spectrum M.

SERPENS. This constellation is divided into parts, Cauda (the body) and Caput (the head), separated by Ophiuchus. Caput is shown in Map IX. The brightest star in Cauda is Eta (3.2); the most interesting object is the fine double Theta, magnitudes 4.5 and 4.5, distance 22", P.A. 103°. This is a splendid object, and is easy to recognize, as it lies in a rather isolated position nor far from Delta Aquilae.

SCUTUM. Though containing no star brighter than the fourth magnitude, Scutum lies in a rich part of the Milky Way, and shows some fine fields. There are several clusters. One of these is the "Wild Duck", M.11, one of the most beautiful open clusters in the sky and shaped like a fan; it lies near Lambda Aquilae. M.26, close to Delta Scuti (4.7), is another good open cluster. It is well worth while to sweep this whole region with low power. R Scuti is an interesting variable.

SAGITTARIUS. This is a large and bright constellation, but is always very low in England and cannot be seen to advantage; part of it never rises at all. The chief stars are Epsilon (1.8). Sigma (2.1, Zeta (2.6), Delta (2.7), Lambda, Pi (2.9), Gamma (3.0), Eta (3.2), and Tau (3.3). Deneb, Altair, and Sagittarius lie almost in a straight line, with Altair in the middle; this is probably the easiest way to find Sagittarius. It can be quite conspicuous on summer evenings. Adjoining Sagittarius, but too far south to be seen in England, is the little constellation CORONA AUSTRALIS (the Southern Crown).
Clusters and Nebulae. M.17: the Omega or Horseshoe Nebula, near Gamma Scuti; a fine object in a moderate telescope.
M.8: the Lagoon Nebula, an easy object near Mu Sagittarii.

M.22: a bright globular between Sigma and Mu, not far from Lambda.

CAPRICORNUS. Like Sagittarius, Capricornus is in the Zodiac. It is a rather barren group; the chief stars are Delta and Beta (each 2.9).

Double Stars. Alpha: magnitudes 3.7, 4.3; distance 376″. This is a naked-eye double, and is easy to find, as the line of stars made up of Gamma Aquilae, Altair, and Beta Aquilae points to it. The fainter component is again double; 3.7, distance 7″; P.A. 158°, and the smaller component of this pair is again double, though a very difficult object.

Beta: a very wide double. Magnitudes 3.1, 6; distance 205″, P.A. 290°. The fainter component is again double; distance 1″.3, P.A. 103°, but the companion is rather faint (10.6) and is thus rather difficult in small apertures.

HERCULES. A small part of Hercules appears in Map VIII, but most of the constellation lies in Map IX. The site of the 1934 nova, DQ Herculis, is marked. This is now a difficult object, and has been found to be a spectroscopic binary. It is described in the text.

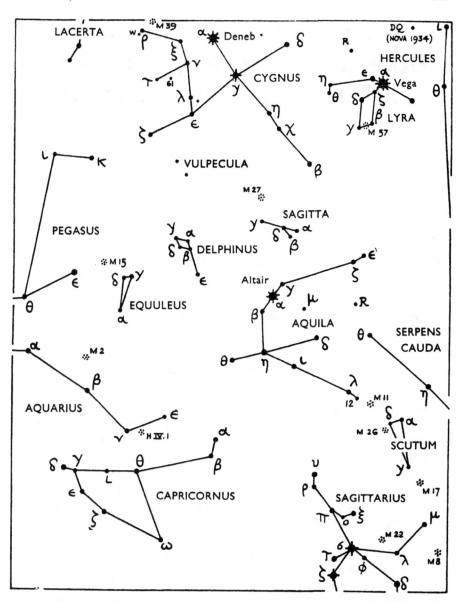

MAP VIII

Map IX. Boötes, Corona Borealis, Hercules, Serpens, Ophiuchus, Libra, Scorpio

These are mainly summer groups, though the northernmost parts of Hercules and Boötes are circumpolar in England. Rough times of rising and setting for Antares, in Scorpio, are as follows:

January 1st	Rises 5 a.m., highest in daylight, sets in daylight.
April 1st	Rises 11 p.m., highest 3 a.m., sets in daylight.
July 1st	Rises in daylight, highest in daylight, sets 1 a.m.
October 1st	Rises in daylight, highest in daylight, sets in daylight.

Arcturus in Boötes is easily recognized, and is shown on Key Map I. Corona is also most conspicuous and can hardly be mistaken. The other groups are less easy to identify, as they are of large area but contain few bright stars. Scorpio is of course an exception, but the most brilliant part of the constellation is always very low in England.

BOÖTES. The chief star is Arcturs; others are Epsilon (2.4), Eta (2.7), Gamma (3.0), and Delta and Beta (3.5). Arcturus is of type K and is distinctly orange.

Double Stars. Epsilon: magnitudes 2.5, 5.3; distance 3″, P.A. 340°. the primary is yellowish, the companion bluish.

Zeta: magnitudes 4.6, 4.7; distance about 1″.3, P.A. 135°. This is close and rather difficult. Binary, period 123 years.

Xi: magnitudes 4.8, 6.9; distance 7″.0, P.A. 344°. Binary, period 152 years.

Delta: magnitudes 3.5, 7.8; distance 105″, P.A. 080°. A very easy object in a small telescope.

Variables. W and R, which lie close to Epsilon. R varies from magnitude 6 to 13 in 225 days; W from 5.2 to 6, irregular.

CORONA BOREALIS. This beautiful little constellation can hardly be mistaken, and it really does look rather like a "crown". The chief stars are Alpha (2.2) and Beta (3.7). Despite its small size, Corona is rich in interesting objects.

Double Stars. Eta: magnitudes 5.7, 5.9; distance 1″, P.A. varies rather quickly, as the star is a binary with a period of 42 years. It is rather close, and is thus not an easy object.

Zeta: magnitudes 4.0, 4.9; distance 6″.3, P.A. 303°. A fine double.

Variables. T: the peculiar nova-like variable. Usually it fluctuates between magnitudes 9 and 10, but it rose to 2 in 1866 and to 3 in 1946. It is well worth watching, as a fresh outburst may occur at any moment.

R: magnitude 5.6 to 14. The well-known irregular variable, described in the text.

S: magnitude 6 to 12, period 361 days. Spectrum M.

HERCULES. A very large but rather barren constellation. It occupies the area between Vega and Corona Borealis. The chief stars are Beta and Zeta (2.8), Alpha (variable), Pi and Delta (3.1), Mu (3.4), and Eta (3.5).

Double Stars. Zeta: magnitudes 3.1, 5.6; distance about 1″. P.A. alters fairly quickly, as the star is a binary with a period of 34 years.

Delta: magnitudes 3.2, 7.5; distance 11″, P.A. 208°.

Alpha: magnitudes 3 (variable), 5.4; distance 4″.6, P.A. 110°. The brighter star is an M5-type giant, reddish; the companion green.

Variables. Alpha. One of the Betelgeux-type irregulars. It fluctuates between magnitudes 3 and 4, and over about 20 years I have found no semblance of a period. The best comparison stars for it are Kappa Ophiuchi (3.42), Gamma Herculis (3.79), and Delta Herculis (3.14).

g: magnitude 4.6 to 6.0. An M-type irregular, near Sigma (4.2).

S: magnitude 6 to 12.5, period 300 days. Spectrum M. It lies between Alpha Herculis and Beta Serpentis.

Clusters. M.13. The famous globular; it lies between Zeta and Eta, and can just be seen with the naked eye under good conditions. It is very easy to find with a telescope, and in a moderate aperture is a glorious sight.

M.92: another globular, between Iota and Eta. It is not unlike M.13, but is far less prominent.

NGC 210 a small bright planetary nebula in the triangle formed by Beta, Delta, and Epsilon Herculis. It is said to have a bluish hue, though to me it always looks white.

OPHIUCHUS. This constellation lies between Vega and Antares. It contains some fairly bright stars: Alpha (2.1), Eta (2.5), Zeta (2.6), Delta (2.7), Beta (2.8), Kappa and Epsilon (3.2), and Mu and Nu (3.3), but it is not easy to identify at first sight, and it is relatively barren of interesting objects. There is a bright globular cluster, M.19, near Theta, and roughly between Theta and Antares; but it is always very low in England. Ophiuchus is not classed as a Zodiacal constellation, but it does enter the Zodiac in the region between Scorpius and Sagittarius.

LIBRA. Zodiacal, but a very dull constellation. The chief stars are Beta (2.6), Alpha (2.8), and Sigma (3.3); Sigma was also included in Scorpius, as Gamma Scorpii. There are few interesting objects apart from the Algol-type eclipsing binary Delta Librae, which has a magnitude range of 4.8 to 6.2 and a period of

2.3 days. Beta Librae is a B8-type star and is said to be the nearest approach to a normal "green" star. It certainly may have a slightly greenish tinge, though the colour is so elusive that many people will fail to detect it. Of course, some double stars have green components, and Nova DQ Herculis was also green at one stage in its career.

SCORPIUS. A splendid Zodiacal group, but never well seen in England; it is always low down, and its "sting" never rises at all. The chief stars, apart from Antares, are Lambda (1.6), Theta (1.9), Epsilon and Delta (2.3), Kappa (2.4), Beta (2.6), Upsilon (2.7), Sigma and Tau (2.8), Pi (2.9), Iota[1] and Mu (3.0), G (3.2), and Eta (3.3), but Lambda, Upsilon, Kappa, Iota, Theta and Eta (3.3) are invisible in England, Regulus and Antares are on roughly opposite sides of Arcturus with Arcturus in the middle, which is of help in identifying Scorpio; Antares is also distinguished by its ruddiness and by the fact that, like Altair, it has a fairly bright star to either side of it – in this case Tau and Sigma Scorpii.

Double Stars. Antares has a companion of magnitude 5.1; distance 3″; P.A. 275°. The primary is of course red; the companion is green. It is a fine object.

Nu: magnitude 4.3, 6.5; distance 41″; P.A. 335°. A wide, easy double. Each component is again double, but very close and difficult.

Beta: magnitudes 2.8, 5.0; distance 1″. There is a third star, magnitude 4.9, at 14″.

Clusters. M.80. A splendid globular, lying roughly between Antares and Beta.

M.4. An open cluster. The stars in it are not brilliant, but the object is not hard to find, as it lies close to Antares.

SERPENS. The chief star in Caput is Alpha (2.6). R is an M-type variable; magnitude 5.7–14.4, period 357 days. M.5, a bright globular, lies near Alpha.

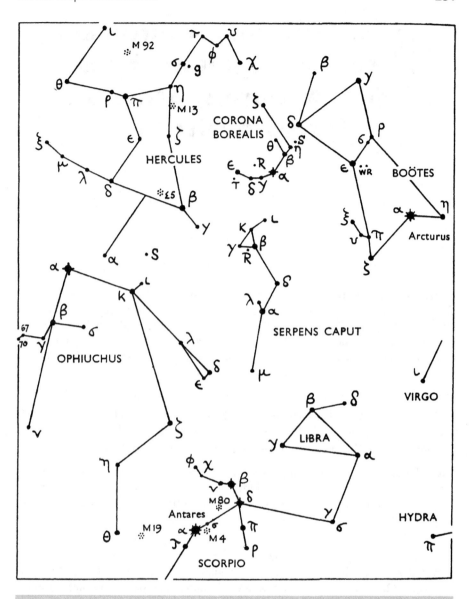

MAP IX

Map X. Pegasus, Andromeda, Pisces, Triangulum, Aries, Cetus, Aquarius, Sculptor, Pisces Australis

The chief group in this map is the Square of Pegasus, which in my view is much more difficult to identify than might be supposed, as most people expect it to be smaller and brighter than it really is. The best way to find it is by means of Cassiopeia, as Gamma and Alpha Cassiopeiae point directly to it. The line from Merak and Dubhe through Polaris will also reach the Square if extended far enough across the sky. Very rough risings and settings are as follows:

January 1st   Rises in daylight, highest in daylight, sets at midnight.

April 1st   Rises 2 a.m., highest in daylight, sets in daylight.

July 1st   Rises in daylight, highest 5 a.m., sets in daylight.

October 1st   Rises in daylight, highest 11 p.m., sets 7 a.m.

It is therefore at its best during the autumn. As is shown on Map VII, one of the stars of the Square is generally included in the neighbouring constellation of Andromeda (Alpha Andromeda=Delta Pegasi). Andromeda and Triangulum are described with Map VII.

PEGASUS. An important constellation, but not so conspicuous as is generally supposed. Alpha Andromedae (2.1) is the Square. The other chief stars of Pegasus are Epsilon (2.3, which is shown on Map VIII), Alpha (2.5), Beta (variable), Gamma (2.8), Eta (2.9), and Zeta (3.4). It is rather instructive to count the number of stars inside the Square visible with the naked eye; there are not very many of them.

Double Star. Xi: magnitude 4.0, 12; distance 12″; P.A. 108°. A difficult double, owing to the faintness of the companion. It is a binary, with a period of about a century and a half.

Variable. Beta: magnitude 2.3 to 2.8. An M-type semi-regular. Suitable comparison stars are Alpha (2.50) and Gamma (2.84). There is a very rough period of about 35 days.

ARIES. Celebrated as being the First Constellation of the Zodiac. It is not, however, very conspicuous. It lies between Aldebaran and the Square of Pegasus and has two fairly bright stars, Alpha (2.0) and Beta (2.7).

Double Star. Gamma: magnitudes 4.7, 4.8; distance 8".2; P.A. 000°. A fine easy double, very well seen with a small telescope. Rather unexpectedly, this is an optical double.

PISCES. The last constellation of the Zodiac, though owing to the precession of the equinoxes it now contains the First Point of Aries. It is large but faint, the brightest star being Eta (3.7). Pisces can be identified by the long line of rather faint stars running below the Square of Pegasus.

Double Stars. Alpha: magnitudes 4.3, 5.3; distance 1".9; P.A. 292°.
Zeta: magnitudes 4.2, 5.3; distance 24"; P.A. 060°.

CETUS. Part of this large constellation is shown in Map IV, and the chief star (Beta) in the key map I. Beta can be found by means of the Square of Pegasus, as Alpha Andromedae and Gamma Pegasi point towards it. Its proper name, Diphda, is often used, and it is an orange star suspected of variability. Not far from it is the M-type semi-regular variable T, which is reddish and has a magnitude range of from 5 to 7. Mira (Omicron), shown here, is described with Map IV.

SCULPTOR (a merciful abbreviation of the old name "Apparatus Sculptoris"). A very obscure constellation near Diphda. It contains no star as bright as the fourth magnitude and no objects of interest to the amateur.

AQUARIUS. Part of this Zodiacal constellation is shown in Map VIII, but most of it lies in the present map. The chief stars are Beta (2.9) and Alpha (3.0); (Map VIII); Delta (3.3) and Zeta (3.7). There is a striking group of orange stars centred around Chi (5.1); these are easy to identify and make pleasing telescopic objects under a low power.

Double Star. Zeta: magnitudes 4.4, 4.6; distance 1".9; P.A. 256°. A fine binary, with a period of 360 years.

Variable. R: magnitude 6 to 11; period 387 days. An M-type long-period variable, not far from the star Omega2 (4.6).

PISCES AUSTRALIS. This small group is also termed Piscis Austrinus. It contains Fomalhaut, of the first magnitude, but no other star as bright as magnitude 4. Fomalhaut can be found by a line drawn from Beta through Alpha Pegasi, in the Square, and continued towards the horizon. From England, Fomalhaut is quite conspicuous near midnight in the autumn months. European observers, however, never see it to advantage. From southern countries it is very prominent and acts as a "guide" to the rather confused area of Southern Birds shown in the map on page 271.

Double Stars. Beta: magnitudes 4.4, 7.8; distance 30"; P.A. 172°.
Gamma: magnitudes 4.5, 8.5; distance 4".3; P.A. 262°.
Delta: magnitudes 4.3, 10.6; distance 5"; P.A. 240°. Rather difficult, owing to the faintness of the companion.

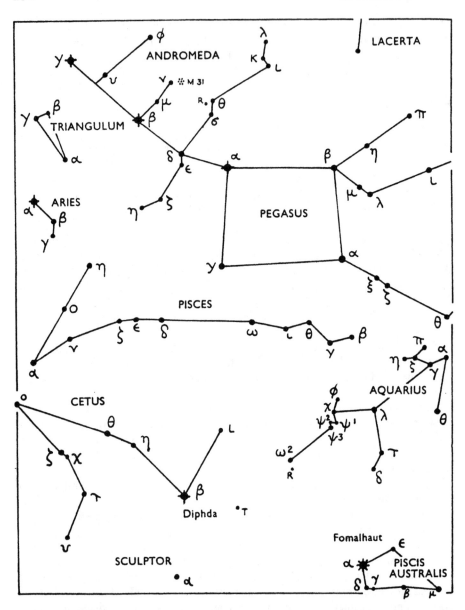

MAP X

The Southern Stars

Map XI. Key Map: Crux Australis (The Southern Cross)

Maps I to X have been drawn for observers who live in the Northern Hemisphere. In fact, the maps are valid for latitudes well south of Europe, with certain modifications; but they do not, of course, apply to countries such as South Africa, Australia or New Zealand. In the Southern Hemisphere, everything is "upside down", and it takes the northern visitor some time to become used to the change. For instance, Leo and Virgo appear inverted, so that the aspect is as shown in this diagram:

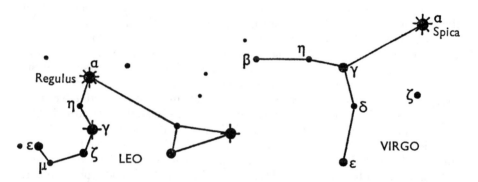

For this book I have not thought it necessary to redraw all the maps, because it is easy to re-orientate them. However, we have not yet dealt with the stars that are too close to the south pole to be seen from Australia. Moreover, we cannot use the Great Bear as a pointer, because it is to all intents and purposes out of view. Instead we have the Southern Cross, which is truly magnificent. We also have Orion, which is cut by the celestial equator and is thus to be seen from all populated parts of the world.

The Southern Cross is the smallest constellation in the whole sky, but it is also one of the most striking. Let us admit that it is not in the least like a cross; it looks rather the shape of a kite. Of its four chief stars, three are brilliant; the fourth, Delta, is considerably fainter, and makes the pattern seem unsymmetrical.

There should be no difficulty in identifying Crux, particularly as the brilliant pair of stars made up of Alpha and Beta Centauri is an ideal guide to it. Alpha Crucis (Acrux) is of the first magnitude; Beta and Gamma are between 1 and $1\frac{1}{2}$. Note, in passing, that Gamma is decidedly red, whereas its neighbours are white.

From South Africa, Crux can drop very low in the sky, as on spring evenings (that is to say, around November; remember that in the Southern Hemisphere summer occurs around Christmas-time and midwinter in June). From Johannesburg, part of it actually sets briefly, though from places further south, such as Cape Town, it scrapes the horizon, and from much of New Zealand it is always to be seen. The maps given in the rest of this section may be regarded as valid for the whole of South Africa, Australia, and New Zealand; the relatively slight latitude variations make no important difference.

The first thing to learn from Crux is the position of the south celestial pole, which lies in a rather blank area. Simply follow the "longer axis" of the Cross, from the red giant Gamma through Acrux. After passing through the polar region, the line will arrive at Achernar, the brilliant leader of Eridanus. Clearly, Achernar and Crux are on opposite sides of the pole, and about the same distance from it – so that when Achernar is high up, Crux will be low down, and vice versa.

Next, follow the "sweep" shown in Map XI. Beta Carinae in the now-dismembered Argo Navis is of magnitude 1.8, and so is bright enough to be conspicuous; beyond it we come to Canopus, which is surpassed only by Sirius. Sirius itself lies well beyond Canopus, too far to be conveniently shown in this key map. Beware of the False Cross, which lies partly in Carina and partly in Vela. In shape it is similar to Crux, but it is much larger, and its stars are not so bright.

Close to Alpha Centauri is Triangulum Australe, the Southern Triangle – one of the few groups to have been given an appropriate name. Alpha, the brightest of the trio, is of above the second magnitude and is distinctly reddish. Follow a line from Alpha Centauri through Alpha Trianguli Australe, and you will eventually come to the second-magnitude star Alpha Pavonis. Beyond lies Grus, the Crane, with its leader Alnair. I have given these stars in the key map because they are the most easily identified objects in an area that is rather confused and undistinctive apart from Grus itself. Alpha Pavonis is circumpolar from Australia, but Grus is not. During winter evenings (southern winter!), it is below the horizon.

Various other groups can be found from Crux. For instance, a line from Acrux through Gamma will show the way to Corvus, so that to all intents and purposes, Achernar, the south pole, Crux, and Corvus are lined up. Acrux and Alpha Centauri act as approximate guides to Antares. And a line from Acrux passed midway between Beta and Gamma Crucis will end up somewhere near Spica in Virgo. Remember that the Y of Virgo is now upside-down by northern reckoning, and the Sickle of Leo curves "down" instead of "up". As soon as Spica and Leo have been found, the other well-known features such as Arcturus and Corona can be located by reorientating the maps given earlier in this section.

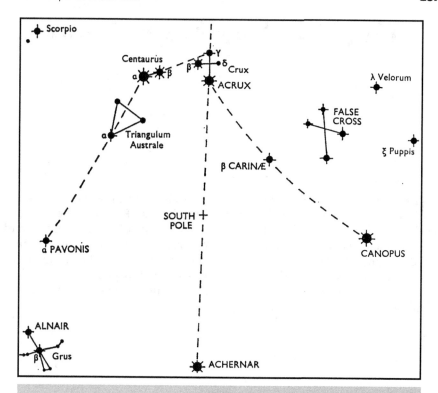

MAP XI

Map XII. Key Map: Orion

Orion is on view for a large part of the year – all through the hot season – and is out of view only during the winter. Now, of course, Rigel is at the "top" and Betelgeux at the "bottom"; the Belt stars point downward to Aldebaran and upward to Sirius. Canopus can be located by using Zeta and Kappa Orionis, and Canopus and Sirius point to the Twins, Castor and Pollux. Regulus and the Sickle can be found by taking a "sweep" from the lower part of Orion through Procyon, as shown in the key map. Canopus is not circumpolar from South Africa or most of Australia, but it spends little time below the horizon.

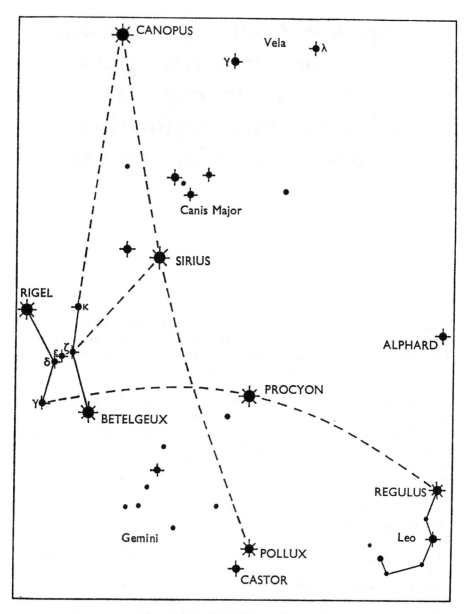

CANOPUS

Vela   λ

Υ

Canis Major

SIRIUS

RIGEL

κ

ALPHARD

δ ε ζ

Υ

PROCYON

BETELGEUX

REGULUS

Gemini   Leo

POLLUX

CASTOR

MAP XII

Map XIII. Crux, Centaurus, Lupus, Circinus, Triangulum Australe, Norma, Ara, Telescopium, Sagittarius, Corona Australis, Scorpio

This map covers a brilliant region of the sky, crossed by the Milky Way. In addition to the Crux-Centaurus area it includes the whole of Sagittarius and most of Scorpius. Both these groups are much more splendid than European dwellers appreciate; when Scorpius is almost overhead it is a magnificent sight.

CRUX is pre-eminent. Its chief stars are Alpha or Acrux (combined magnitude 0.9), Beta (1.3), Gamma (1.6), and Delta (3.1); there is also Epsilon (3.6), which tends to upset the pattern. Though there is so little difference between Beta and Gamma, Beta is unofficially ranked as of the "first magnitude", whereas Gamma is not. Binoculars will show a striking difference between Gamma, a red giant of spectral type M, and the remaining members of the Cross, all of which are hot and white.

Double Stars. Alpha (Acrux): magnitudes 1.6, 2.1; distance 4".7; P.A. 114°. This is a splendid double, separable with a small telescope (it is very easy in a 2-inch); there is also a third star in the field.

Gamma: magnitudes 1.6, 6.7; distance 110".6; P.A. 031°. A wide optical double.

Iota: magnitudes 4.7, 7.8'; distance 26".4; P.A. 027°. Very easy.

Mu: magnitudes 4.3, 5.5; distance 34".9; P.A. 017°. Also very easy.

Variables. R: magnitude 6 to 8; period 5.8 days; a Cepheid.

S: magnitudes 6.6 to 7.7; period 4.7 days; also a Cepheid.

T: magnitudes 6.9 to 7.7; period 6.7 days; yet another Cepheid. All these three Cepheids may be followed with binoculars.

Clusters and Nebulae. Kappa Crucis, the so-called Jewel Box, is a superb loose cluster in which there are stars of different colours. It is easily identifiable in binoculars, and I have no hesitation in calling it the most glorious cluster in the whole sky. Close to it is the celebrated dark nebula called the Coal Sack, again visible in binoculars. There are a few foreground stars, but the dark mass is really striking.

Crux may be the smallest of all the constellations, but it is amazingly rich in interesting objects.

CENTAURUS. This is another really splendid constellation. Its leading stars are Alpha (combined magnitude –0.3), Beta (Agena) (0.7), Theta (2.2), Gamma (2.3), Eta (2.6), Epsilon (2.6), Iota (2.9), Delta (2.9), Zeta (3.0), Kappa (3.3), Mu (3.3), and Lambda (3.3). Alpha Centauri is striking; it is surprising that it has no old-established official name. It makes up a magnificent pair with Beta or Agena. Centaurus has a distinctive shape, and more or less surrounds Crux.

Double Stars. Alpha: magnitudes 0.3, 1.7. This is a superb binary with a period of 80 years. Both position angle and distance alter fairly rapidly, but the average separation is about 4″, so that the pair is easily split in a small telescope. Beta has a ninth-magnitude companion at P.A. 255°, but the distance is only 1″.4. Gamma is a close binary; the components are almost equal (magnitudes 3.1, 3.2) and the period is 84½ years.

Variables. R: magnitude 5.4 to 11.8; period 547 days. A typical long-period variable of the Mira type.

T: magnitude 5.5 to 9.0; period 91 days. This is classed as a semi-regular variable.

Clusters and Nebulae. Omega Centauri is much the finest globular cluster in the sky. To the naked eye it appears as a hazy star of the fourth magnitude; binoculars show it well, and a fairly small telescope will resolve it.

NGC 3766. A fine cluster near Lambda, visible with binoculars.

CIRCINUS. A small constellation between Alpha Centauri and Triangulum Australe. Its leading stars are Alpha (3.4) and Beta (4.2). Alpha is a double; magnitudes 3.4, 8.8; distance 15″.8; P.A. 235°. The primary is of spectral type F, and is distinctly yellowish.

NORMA. A very obscure constellation; the brightest star, Gamma², is only of magnitude 4.1. The only object of note is the open cluster NGC 6067, which is 20' in diameter, but is not particularly conspicuous.

TRIANGULUM AUSTRALE. A prominent triangle, made up of Alpha (1.9) and Beta and Gamma (each 3.1). Alpha is obviously reddish.

Variable. S: magnitude 6.4 to 7.6; period 6.3 days. A Cepheid.

Cluster. NGC 6025. A bright open cluster, visible in binoculars.

ARA. A fairly prominent constellation; the leading stars are Beta (2.8), Alpha (3.0), Zeta (3.1), Eta (3.7), Delta (3.8), and Theta (3.9). There are no notable objects except for two Algol variables, R (6.0 to 6.9, 4.4 days) and RW (8.7 to 12, also 4.4 days).

LUPUS. This is a decidedly "shapeless" constellation, adjoining Centaurus. It contains some moderately bright stars, of which the chief are Beta (2.8), Alpha (2.9), Gamma (2.9), Delta (3.4), Zeta (3.5), and Phi¹ (3.6).

Double Stars. Kappa: magnitudes 4.1, 6.0; distance 27″; P.A. 144°. Very wide and easy.

Eta: magnitudes 3.6, 7.7; distance 15″.2; P.A. 021°. Also easy.

Pi: magnitudes 4.7, 4.8; distance 1″.7; P.A. 076°.

Mu: magnitudes 5.0, 5.2; distance 1″.4: P.A. 146°.

TELESCOPIUM. A small constellation adjoining Ara. The leading stars are Alpha (3.8) and Zeta (4.1). It contains nothing of note.

CORONA AUSTRALIS. A small semicircle of stars, of which the brightest are Alpha (4.1) and Beta (4.2). Gamma is double: magnitudes 5.0, 5.1; distance 2″.7; P.A. 054°. Though its stars are faint, Corona Australis is easy to recognize, and is worthy of separate identity instead of being included in Sagittarius.

Also on this map are parts of SCORPIUS and OPHIUCHUS and all of SAGITTARIUS. The southernmost parts of Scorpius and Sagittarius are not easily seen from Europe – which is a pity, because Scorpio at least is a splendid constellation. The "sting", made up of Lambda (1.7), Kappa (2.5), Upsilon (2.7), Iota1 (3.1), and G (3.2), is very distinctive, and close by lies Theta (2.0). The two open clusters M6 and M7 are worth studying. The brightest star in Sagittarius, Epsilon (1.8), lies not far from the Scorpion's sting. Observers who live in Europe or North America cannot be expected to appreciate how brilliant and distinctive Scorpio and Sagittarius really are.

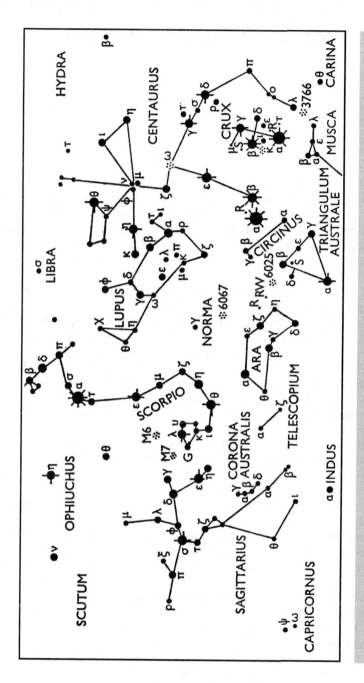

MAP XIII

Map XIV. Carina, Vela, Puppis, Pyxis, Volans, Antila

This map is occupied the old Argo Navis, which has now been divided up into smaller constellations. Crux, Beta Carinae, Canopus, and Sirius form a magnificent curved line that cannot be misidentified. The Falso Cross is made up of Epsilon and Iota Carinae, and Kappa and Delta Velorum. The whole region is exceptionally rich and is crossed by the Milky Way.

CARINA. This contains Canopus, which is inferior only to Sirius. Canopus has an F-type spectrum, and is usually described as yellow, though I admit that to me it always looks colourless. Its magnitude is –0.7. The other leading stars of Carina and Epsilon (1.7), Beta (1.8), Iota (2.2), Theta (3.0), Upsilon, Omega (3.6), Chi (3.6), and p (3.6). Most of them are hot and white, but Epsilon is a beautiful orange star of type K. The small constellation of Volans intrudes into Carina.

Double Star. Upsilon: magnitudes 3.1, 6.0; distance 4″.6; P.A. 126°.

Variables. Eta: the most erratic of all variables, once the rival of Canopus and now below magnitude 6. It lies in a particularly rich area and may at any time regain prominence.

R: magnitude 3.9 to 10.0; period 308.6 days. A Mira-type variable.

S: magnitude 4.5 to 9.9; period 149.5 days. Also of the Mira type.

I: magnitude 3.6 to 5.0; period 35.5 days. A Cepheid. Its long period (by Cepheid standards) indicates very high luminosity.

U: magnitude 6.4 to 8.4; period 38.8 days. Another Cepheid, also with a period of unusual length for a star of its type.

Clusters and Nebulae. Carina is well supplied with rich fields, and the nebulosity associated with Eta is worthy of special note. NGC 2516, near Epsilon, is a rich open cluster visible with the naked eye; NGC 2808 is a good example of a globular.

VELA. Principal stars: Gamma (1.9), Delta (2.0), Lambda (2.2), Kappa (2.6), Psi (3.6), and c, Phi, and Omicron (each 3.7). Gamma is the brightest of all the Wolf-Rayet stars. Kappa and Delta make up part of the False Cross, together with Epsilon and Iota Carinae.

Double Stars. Gamma: magnitudes 2.2, 4.8; distance 41″; P.A. 220°. Very wide and easy. Each component is itself a spectroscopic binary.

Delta: magnitudes 2.0, 6.5; distance 2″.9; P.A. 164°. Easy.

Mu: magnitudes 2.8, 7.0; distance 1″.0; P.A. 079°.

Psi: magnitudes 4.2, 4.7. P.A. and distance alter quickly, as this is a binary with the relatively short period of 34.1 years. The mean separation is about 0″.4, so that this is too close a pair to be easy.

Variables. There are two short-period variables within binocular range. AH is a Cepheid (magnitude 5.8 to 6.4; period 4.2 days), while AI is an RR Lyrae type star with a range of 6.4 to 7.1 and a period of only 0.11 days. In addition, N Velorum, which is of type K5 and is distinctly reddish, is a suspected variable well worth watching; it lies close to the False Cross. Its official magnitude is rather above 5.

Clusters. There are several clusters in Vela, though none is of special note. The whole area is extremely rich, and is well worth sweeping with binoculars.

PUPPIS. This is the northernmost part of the old Argo, and part of it is visible from European latitudes. The principal stars are Zeta (2.3), Pi (2.7), Rho (2.9), Sigma (3.2), Nu (3.2), and Xi (3.5). Zeta is another very hot star of spectral type O.

Double Stars. Sigma: magnitudes 3.2, 8.5; distance 22″.4; P.A. 074°.

Xi: magnitudes 3.5, 13.5; distance 4″.3; P.A. 189°. I include this as an example of a really difficult object, in view of the faintness of the companion.

Variables. L^2. This is an ideal binocular object. The range is from 2.6 to 6.0, but it is a semi-regular variable (period 141 days) and the light-curve never repeats itself exactly. It is a red giant of type M. Suitable comparison stars are L^1 (magnitude 5.0), C (5.3), I (4.5), and Sigma (3.2).

V: magnitude 4.5 to 5.1; period 1.45 days. An eclipsing binary of the Beta Lyrae type, near Gamma Velorum.

Z: magnitude 7.2 to 14.6; period 510 days; Mira type.

Cluster. M.46; in the northern part of Argo, and described on page 277.

PYXIS. A small group, with only Alpha (3.7) above the fourth magnitude. The only object of interest is T, a recurrent nova that is usually of about the 14th magnitude, but which increased to 7.0 in 1920 and again in 1944.

ANTLIA. An even less remarkable constellation. The brightest star is Alpha (magnitude 4.4), and there are no noteworthy objects.

VOLANS. A constellation that intrudes into Carina, near Beta. The chief stars are Beta (3.7), Zeta (3.9), Gamma2 (3.9), Delta (4.0), and Alpha (4.2).

Double Stars. Gamma: magnitudes 3.9, 5.8; distance 13″.8; P.A. 299°. Very wide and easy.

Epsilon: magnitudes 4.5, 8.0; distance 6″.1; P.A. 022°.

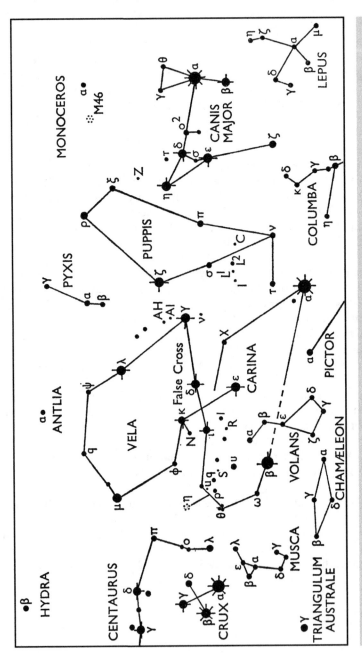

MAP XIV

Map XV. Octans, Apus, Musca, Chamaeleon, Mensa, Hydrus, Reticulum, Dorado, Pictor, Eridanus, Horologium

This is the south polar area. It is divided up into a number of relatively dim constellations. Broadly speaking, the region is enclosed by imaginary lines connecting Canopus, Achernar, Alpha Pavonis, and Alpha Trianguli Australe. As has been shown on the key map, the pole may be located by using the longer axis as a guide.

OCTANS. A remarkably barren constellation. The brightest star is Nu (magnitude 7); Sigma, which lies close to the pole, is only of magnitude 5.5. The only object worthy of mention is the long-period variable R (magnitudes 6.4 to 13.2; period 405 days, Mira type).

APUS. Fairly compact; the leading stars are Alpha (3.8) and Gamma (3.9).
Variable. Theta; magnitudes 5.1 to 6.6. The spectrum is of type M, so that the star is red. It seems to be genuinely irregular in behaviour.

MUSCA. This is a conspicuous little constellation adjoining Crux. It contains some fairly bright stars, closely grouped. The leaders are Alpha (2.9), Beta (3.2), Delta (3.6), Lambda (3.8), and Gamma (4.0).
Double Stars. Beta; magnitudes 3.9, 4.2; distance 1".6; P.A. 007°. A good example of a pair with almost equal components.
Theta: magnitudes 5.6, 7.2; distance 5".7; P.A. 186°.
Variables. R: magnitude 6.3 to 7.3; period 7.5 days. A Cepheid.
S: magnitude 6.2 to 7.3; period 9.7 days. Another Cepheid.

CHAMAELEON. Very obscure, with no stars brighter than Alpha and Gamma (magnitude 4.1). Delta is made up of a pair of stars, of magnitudes 4.6 and 5.5, respectively; the brighter is white, the fainter orange. However, the separation is too wide for classification as a double, and the two are not genuinely associated with each other.

MENSA. This would be one of the most unremarkable of all constellations but for the presence of the Large Cloud of Magellan. The brightest star in Mensa (Gamma) is only of magnitude 5.1, but the Cloud is superb; it has been described in the text. Binoculars show it excellently, but it is of course a prominent naked-eye feature. It extends from Mensa into Dorado.

HYDRUS. Chief stars: Beta (2.9), Alpha (3.0), and Gamma (3.1). Alpha lies close to Achernar. Though it has three fairly bright stars, Hydrus is remarkably deficient in interesting objects.

RETICULUM. Another group that is compact enough to be easily identifiable; its leaders are Alpha (3.4) and Beta (3.8).
Double Star. Theta: magnitudes 6.2, 8.3; distance 3″.9; P.A. 004°.
Variable. R: magnitude 6.8 to 14.0; period 278 days. Mira type.

DORADO. A constellation notable chiefly for containing part of the Large Cloud of Magellan. Its brightest stars are Alpha (3.5) and Beta (a variable; at maximum, 3.8).
Variable. Beta: magnitude 3.8 to 5.0; period 9.8 days; a Cepheid.
Nebula. The Large Cloud has already been described, but mention should be made here of the Looped Nebula, 30 Doradûs, which is visible to the naked eye in the Cloud. With any optical aid it is a superb sight.

PICTOR. A constellation between Dorado and Canopus; its leader is Alpha (3.3).
Double Star. Iota: magnitudes 5.6, 6.4; distance 12″: P.A. 058°.
Variables. R: magnitude 6.7 to 10.0; period 171 days. Semi-regular.
RR. In 1925 the bright nova RR Pictoris flared up here. It became very prominent but is now extremely faint, and there is no prospect of its undergoing a second outburst.

ERIDANUS. Part of Eridanus is shown in this map; of course the most brilliant star, Achernar, is much too far south to be seen from Europe. Also in the southern part of the constellation is Theta, a splendid double; magnitudes 3.4, 4.4; distance 8″.5; P.A. 088°. To the naked eye Theta appears of magnitude 3.1, but the ancient observers ranked it as of the first magnitude, and it may have faded since then, though the evidence is very far from conclusive. Both components are of spectral type A2.

HOROLOGIUM. A very obscure group adjoining Eridanus; its only moderately bright star is Alpha (3.8). In it is the long-period variable R, which varies between magnitudes 4.7 and 14.3 in 402.7 days and is of the Mira type.

Also included in this map are Phœnix, Tucana, and Pavo, but these are best described with Map XVI. Note that the Small Cloud of Magellan adjoins Hydrus and actually extends into it, though most of it lies in Tucana.

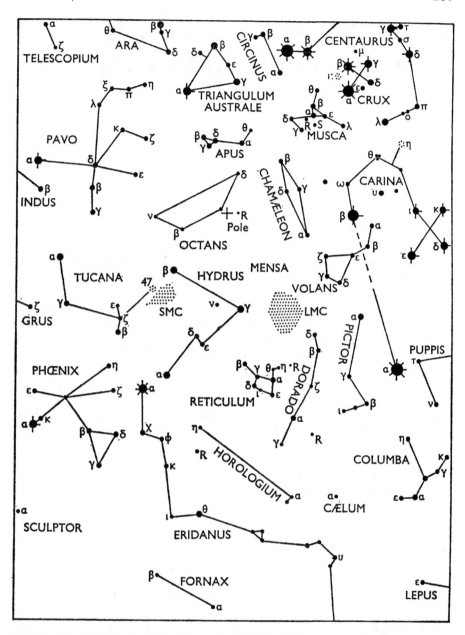

MAP XV

Map XVI. Pavo, Indus, Tucana, Grus, Phoenix, Microscopium

This is the region of the "Southern Birds". I have found that the best means of identification is to locate Alpha Pavonis by using Alpha Centauri and Alpha Trianguli Australe, as shown in the key map. Of all the groups, only Grus is distinctive.

PAVO. The brightest star is Alpha (2.1); then follow Delta, Eta and Beta (each 3.6), and Kappa (4.0 at maximum). Alpha is rather isolated from the rest of the constellation. Delta is 19 light-years away and is very like the Sun in every respect; it is interesting to speculate whether it has a similar system of planets!

Double Star. Xi: magnitudes 4.3, 8.6; distance 3".3; P.A. 151°.

Variables. Kappa; magnitude 4.0 to 5.5; period 9.1 days. W Virginis type. Type 2 Cepheid.

R, S, T: All these are of the Mira type. R: magnitudes 7.5 to 13.8; period 230 days. S: magnitudes 6.6 to 10.4; period 387 days. T: magnitudes 7.0 to 14.0; period 244 days.

INDUS. A group near Alpha Pavonis. Its leaders are Alpha (3.2) and Beta (3.7). The only object of note is the double star Theta; magnitudes 4.6, 7.0; distance 5".3; P.A. 276°.

TUCANA. The Toucan is enriched by the presence of the Small Cloud of Magellan as well as the superb cluster 47 Tucanae. Chief stars: Alpha (2.9), Beta (3.7), Gamma (4.1), and Zeta (4.3). Alpha Tucanae, Alpha Pavonis, and Alnair (Alpha Gruis) form a triangle.

Double Stars. Beta: magnitudes 4.5 and 4.5, giving a combined naked-eye magnitude of 3.7; distance 27".1, P.A. 170°. A superb easy pair. Each component is again double, though in each case the separation is small. In the same field is yet another star, of magnitude 5, which is itself double. The group is very well worth careful study.

Delta: magnitudes 4.8, 9.3; distance 6".8; P.A. 283°.

Kappa: magnitudes 5.1, 7.3; distance 5".7; P.A. 341°.

Clusters and Nebulae. The Small Cloud lies almost entirely in Tucana, and it too is a prominent naked-eye object, though moonlight will drown it. On its fringes are two splendid globulars. 47 Tucanae is surpassed only by Omega Centauri, as it is easy to see with the naked eye and is magnificent with any optical aid. Also close to the Cloud is NGC 362, which is just visible with the naked eye and has a diameter of 10 minutes of arc.

GRUS. A very prominent and distinctive constellation that really does give some impression of a flying crane! Its chief stars are Alpha or Alnair (2.1), Beta (2.2), Gamma (3.1), and Epsilon (3.7). Alnair and Beta make a good contrast because Alnair is white and Beta is orange-red. In the line of stars extending from Beta to Gamma are two pairs, Delta[1] and Delta[2], and Mu[1] and Mu[2]. Both are easy to separate with the naked eye, and are too wide to be classed as bona-fide doubles.

Double Star. Theta: magnitudes 4.5, 7.0; distance 1".5; P.A. 052°.

Variables. R: magnitude 7.4 to 14.9; period 332.5 days; Mira type.

S: magnitude 6.0 to 15; period 401 days; also Mira type.

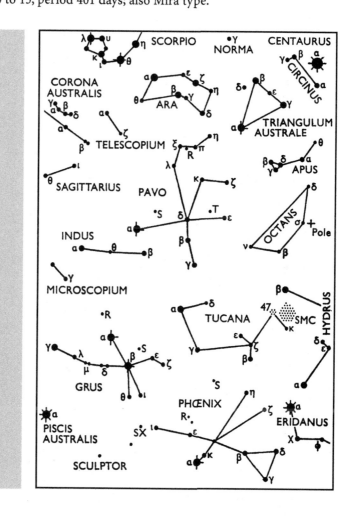

MAP XVI

PHOENIX. A less obvious group, extending to the region between Grus and Achernar. It has one bright star, Alpha or Ankaa (2.4); then follow Beta (3.3), Gamma (3.4), and Delta (4.0).

Double Stars. Beta: magnitudes 4.1, 4.1; distance 1″.3; P.A. 352°.

Zeta: magnitudes 4 (variable) and 7.2; distance 0″.8; P.A. 039°.

Eta: magnitudes 4.5, 11.4; distance 19″.8; P.A. 216°. Not easy, because of the faintness of the companion.

Variables. Zeta: magnitude 3.6 to 4.1; period 1.67 days. An eclipsing binary.

R: magnitudes 7.5 to 14.4; period 268 days. Mira type.

S: magnitude 7.4 to 8.2; period 141 days. Semi-regular.

SX: magnitude 6.5 to 7.5; period 0.055 days. RR Lyrae type.

MICRCOSCOPIUM. An entirely obscure constellation, adjoining Indus.

Appendix 26

The Observation of Variable Stars

The observation of variable stars is becoming more and more popular among amateurs. Some notes on it have already been given (Chapter 15). To present a full account would need a complete book to itself, particularly as there are so many variables within range of a small telescope; clearly this is impossible here, but I can at least provide some "typical cases" to suit various types of equipment. I have done my best to give a general survey, though I admit to having confined myself in the main to stars that are on my own observational list. First, however, it may be as well to summarize the various classes:

Eclipsing variables. These, as we have seen, are not true variables at all, but are binary systems. The main types are:

1. Algol. One component much brighter than the other, producing one marked minimum and a second minimum that is too small to be noticeable.
2. Beta Lyrae. Components less unequal, and very close together; both minima noticeable.
3. W Ursae Majoris. Close binaries; components often about equal to the Sun; short periods, often less than 12 hours. No bright examples.

To study eclipsing binaries properly needs photoelectric equipment, which few amateurs will have. In fact, there are some stars in which visual work is useful; but they are rare, and I have never tackled them myself, which is an extra reason for saying no more about them here.

Pulsating Variables

(a) Short Period

1. RR Lyrae stars. Very regular; very short periods; common in globular clusters, though many of them (including RR Lyrae itself) are not cluster members. No bright examples. All RR Lyrae stars are of approximately the same luminosity, so that they act as distance markers.

2. Cepheids, such as Delta Cephei, Eta Aquilae, and Beta Doradûs, already described in the text. Some have a considerable range in magnitude; others change very little. For instance, Polaris is a Cepheid, but its range amounts to less than 0.2 magnitude. The changes are very regular, and again photoelectric equipment is needed. Classical Cepheids belong to Population I.

3. W Virginis stars. These are Population II Cepheids, with a rather different period-luminosity law. About 50 are known, but none is brilliant in our skies. Again, photoelectric equipment is needed.

There are various other classes of regular short-period variables, but I do not propose to discuss them here, because they are not suited to amateur observation.

(b) Longer Period

1. Mira-type stars, such as Mira Ceti, R Cygni, Chi Cygni, and U Orionis. Both period and range alter, and the light-curve is never repeated exactly from one cycle to the next. Most of them are red giants. Because they are unpredictable, they are ideal amateur objects; it is quite good enough to estimate their magnitudes down to 0.1.

2. Semi-regular stars, such as R Lyrae. Smaller ranges, and periods that are generally shorter; but the periods are very rough indeed and are subject to interruption. Amateur observation of them is very valuable.

3. R V Tauri stars. Alternate deep and shallow minima, but the light-curves are never repeated exactly, and the behaviour is often quite irregular for a while. R Scuti is the brightest example.

(c) Irregular

This is a general term; some stars that are classed as irregular may in fact be semi-regular, but insufficiently observed. A splendid example is Rho Cassiopeia. Mu Cephei, Herschel's "Garnet Star", also seems to be quite irregular.

Eruptive Variables

1. SS Cygni or U Geminorum stars. Nova-like outbursts at mean intervals that range from 20 to 600 days, but which are never predictable. Ideal for amateur observation, but most of them are rather faint, and large apertures are needed.

2. R Coronae stars. These remain at or near maximum for most of the time, but exhibit sudden, unpredictable drops to minimum. They contain more than their fair share of carbon, but are deficient in hydrogen. Observation of them is very useful. Only R Coronae itself is ever visible with the naked eye; stars of this type are rare, and most of them are inconveniently faint.

3. T Tauris stars. Rapid, irregular fluctuations. These seem to be very young stars, but most are faint; T Tauri itself is the brightest of its class (magnitude about 9).

4. A Camelopardalis stars. Similar to SS Cygni stars except that at unpredictable intervals the fluctuations cease for a while, and there is a "standstill". Rare; large apertures needed.

5. Flare stars, such as UV Ceti and AD Leonis. These show sudden rises amounting perhaps to several magnitudes; the outburst takes only a few minutes, and the subsequent fading may take hours. Again, most of them are faint, and the observation technique is different from that used for other variables; the star is kept under constant observation for a set period. I have spent many tens of hours in observing them, but I have only seen one "perform". This was AD Leonis, which is easy to find because it lies in the field with Gamma Leonis. Its normal magnitude is 9.5, but it can flare up to above 9. UV Ceti, the prototype, is usually 12.9, but on one occasion was seen to increase to 5.9! All these stars are nearby red dwarfs. Observation of them is fascinating, but is a matter for the real specialist with endless patience.

6. P Cygni stars. Slow, erratic variations; may be related to novae. The only bright example is P Cygni itself, which is sometimes classed as a nova (1600) but since about 1715 has remained of about the fifth magnitude. All are extremely hot, luminous and remote.

7. Novae. Rapid rise, followed by a slower decline. The outstanding examples of recent years have been HR Delphini (1967) and Nova Cygni (1975).

When estimating the magnitude of a variable, it is essential to use several comparison stars. Either the step-method or the fractional may be employed. (I use the step, though many people tell me that the fractional is actually rather more accurate.) What usually happens, of course, is that a discrepancy is found. Suppose you estimate the variable as being 0.3 below comparison star A, and 0.5 above B; on looking up your charts you find that A is of magnitude 7.0, B is 7.6. From A, the variable would work out at 7.3; from B, 7.1. By using three or more comparison stars a good figure can usually be obtained, but odd things can happen sometimes; on more than one occasion a comparison star has been found to be itself variable, which leads to very peculiar results! Unfortunately, it is not easy to compare a red star with a white one, and many long-period variables are red. U Cygni, which is intensely red, is notoriously difficult to estimate correctly, as I know to my cost.

The following notes and charts are specimens only. Anyone who is interested can obtain others; if he belongs to the British Astronomical Association or the American Association of Variable Star Observers, there will be no difficulty in this respect. Binoculars variables are dealt with on pages 276 to 279 and telescopic variables on pages 280 to 284.

Binocular Variables

Many interesting variables are within the range of binoculars, and there are a few which can be estimated with the naked eye – though extinction must always be allowed for. Naked-eye variables are Betelgeux, Alpha and Gamma Cassiopeiae, Alpha Herculis, Delta Cephei, Kappa Pavonis, Beta Doradûs, and various others; the comparison stars can be looked up from the maps and notes in the previous section.

If only one pair of binoculars is available, a good pair may be 7 × 50 (magnification 7; diameter of each O.G., 50 mm). With magnification of over 12 or so, it is a good idea to have a mounting, which can easily be made. For my 20 × 70 binoculars, I made a stand out of a plank and broomhandles, which is rudimentary, but which works well.

R Lyrae. Semi-regular. 4.0 to 5.0. Comparison stars, Eta and Theta (4.5) and 16 (5.1). The period is said to be about 46 days. An awkward star, because there are no suitable comparisons close to it; but it is very easy to find. Do not be surprised if your light-curve seems odd!

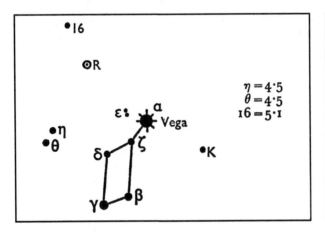

$\eta = 4\cdot5$
$\theta = 4\cdot5$
$16 = 5\cdot1$

R LYRAE Naked-eye view

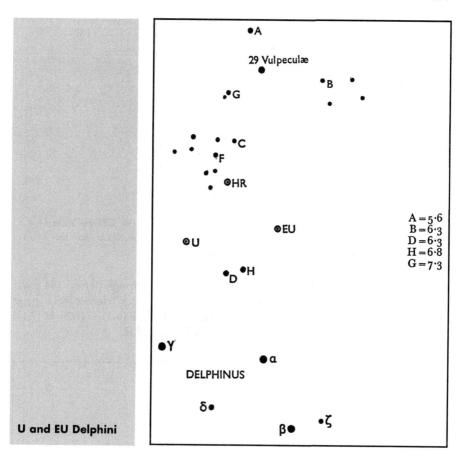

U and EU Delphini

29 Vulpeculæ

A = 5·6
B = 6·3
D = 6·3
H = 6·8
G = 7·3

DELPHINUS

U and EU Delphini. U has a range of 5.6 to 7.5; irregular. EU ranges from 6.0 to 6.9; semi-regular, period around 60 days. Comparisons: A (5.6, but inconveniently far away); B (6.3), D (6.3); H (6.8); and G (7.3). HR, the 1967 nova, is now too faint for binocular use.

W Cygni. A very interesting star. The range is 5.0 to 7.6, and it is officially classed as semi-regular with a period of 130 days, though my own observations since 1968 certainly do not confirm this. To locate it, find Rho Cygni (map, page 247). The field will instantly be recognizable in binoculars. Comparisons: 75 Cygni (5.0, D (5.4), A (6.1), B (6.7), K (6.8), L (7.5). Beware of the star marked X, which is itself variable by half a magnitude. (Note that the letters in these charts are those used by variable star observers, and are not "official"; for instance, X Cygni, near Lambda, is a Cepheid, and is not the same as the X on this chart). Note 75 Cygni, I shall return to it later, as it is the guide to another famous variable, SS Cygni.

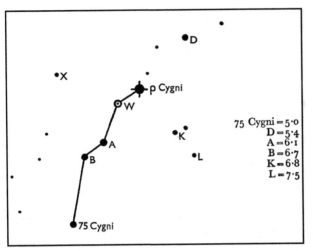

75 Cygni = 5·0
D = 5·4
A = 6·1
B = 6·7
K = 6·8
L = 7·5

W CYGNI Naked-eye or binocular view

R Scuti. The brightest of the R V Tauri stars. Range 5 to 7; rough period 144 days. It is easily found, Lambda Aquilae and the famous "Wild Duck" cluster M.11 (page 247). Comparisons: A (4.5), B (4.8), C (5.0), D (5.2), E (5.6), F (6.1), G (6.8), H (7.1), K (7.7). R makes a well-marked quadrilateral with F, G and H.

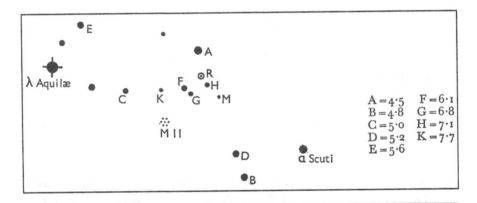

A = 4·5   F = 6·1
B = 4·8   G = 6·8
C = 5·0   H = 7·1
D = 5·2   K = 7·7
E = 5·6

R SCUTI Naked-eye or binocular view

Rho Cassiopeiae. Close to Beta (page 242). An ideal binocular object. Its usual magnitude is about 5, but its official range is 4.1 to 6.2; its drops to minimum are infrequent (I have never seen one yet), and nobody knows what sort of variable it is. It is a very remote super-giant. Comparisons: Theta (4.5; page 242), Sigma (4.9), Tau (5.1), H (5.7), K (6.1).

These are only half a dozen of the many variables which may be followed with binoculars. Also, some long-period variables, such as R Leonis, U Orionis and R Serpentis, are binocular objects when near maximum; and Mira Ceti, of course, is a naked-eye object when at its best. In 1969 it even approached the second magnitude.

RHO CASSIOPEIAE Naked-eye or binocular view

Telescopic Variables

Again I give only a few specimen examples and charts. There are so many variables within range of even a 6-inch telescope that no single observer can hope to deal with them all, and one has to make out a personal list. My own (2004) included 51 stars, which is as many as I could manage; others will certainly be able to do better. The following charts are inverted, for telescopic use.

R Cygni. This is extremely easy to find, since it lies in the field with the 4th-magnitude star Theta Cygni – just off the map on page 240, but shown here. It has a range of from 6.5 to 14.2, and a period of 426 days. At its brightest it is extremely easy, and is visible in binoculars; at minimum it needs a large aperture.

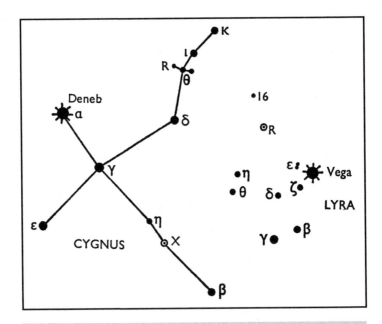

LOCATING θ CYGNI (guide star for R Cygni) Naked-eye or binocular view

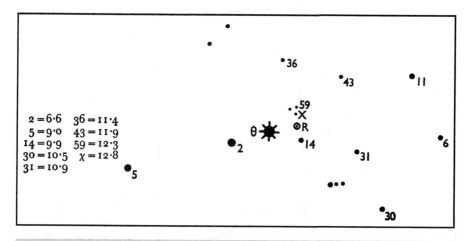

$$
\begin{array}{ll}
2 = 6 \cdot 6 & 36 = 11 \cdot 4 \\
5 = 9 \cdot 0 & 43 = 11 \cdot 9 \\
14 = 9 \cdot 9 & 59 = 12 \cdot 3 \\
30 = 10 \cdot 5 & \chi = 12 \cdot 8 \\
31 = 10 \cdot 9 &
\end{array}
$$

FIELD FOR R CYGNI Telescopic view

Theta Cygni is easily found; near it is the comparison star 2 (6.6). Other comparison stars are 5 (9.0), 14 (9.9), 31 (11.0), 36 (11.4), 43 (11.9), 59 (12.3) and × (12.8). This is not a full sequence, but it will be enough to show the way in which R Cygni behaves. It is of type S, and very red. Of course, it passes below the range of small telescopes when faint.

SS Cygni. This is an excellent example of a fainter variable which is easy to find. Locate 75 Cygni, near Rho, as already described; it is identifiable because it is distinctly red. Then look for the triangle made up of C (8.5), G (9.6) and F (9.4); SS lies between C and G. Also to hand are A (8.0), N and 49 (each 11.3), O (11.8) and P (12.1). At its usual brightness SS is comparable with O; at its best it can equal A. The average period between outbursts is 50 days, but this *is* only an average. Apparently all SS Cygni stars are spectroscopic binaries.

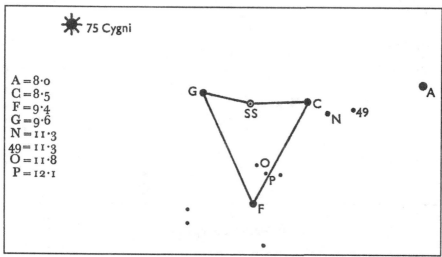

$$
\begin{array}{l}
A = 8 \cdot 0 \\
C = 8 \cdot 5 \\
F = 9 \cdot 4 \\
G = 9 \cdot 6 \\
N = 11 \cdot 3 \\
49 = 11 \cdot 3 \\
O = 11 \cdot 8 \\
P = 12 \cdot 1
\end{array}
$$

SS CYGNI Telescopic view

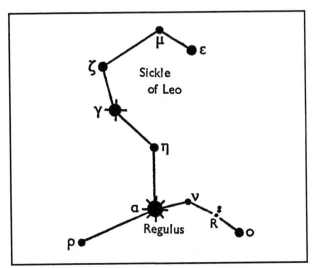

LOCATING R LEONIS Naked-eye view

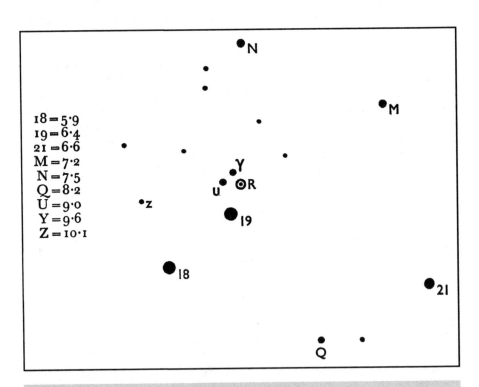

18 = 5·9
19 = 6·4
21 = 6·6
M = 7·2
N = 7·5
Q = 8·2
U = 9·0
Y = 9·6
Z = 10·1

R LEONIS Telescopic view

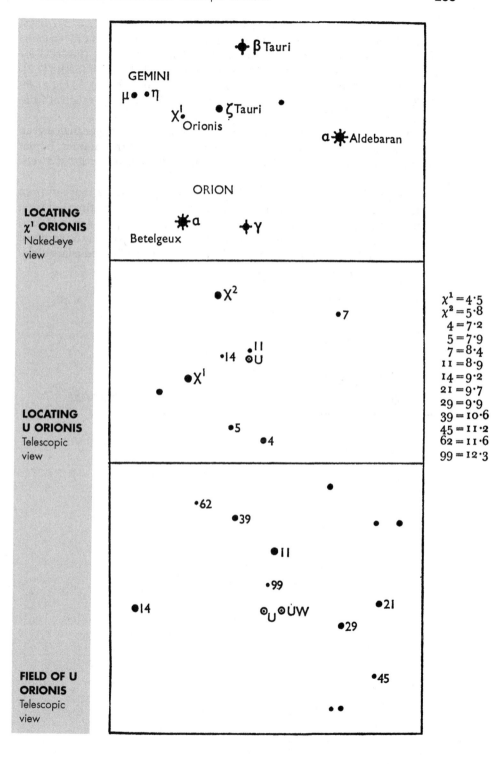

LOCATING
χ¹ ORIONIS
Naked-eye
view

LOCATING
U ORIONIS
Telescopic
view

FIELD OF U
ORIONIS
Telescopic
view

β Tauri

GEMINI
μ• •η

χ¹
Orionis   •ζ Tauri

α☀Aldebaran

ORION

Betelgeux   ☀α   ✦γ

•χ²   •7

•14   •11
⊙U

•χ¹   •5   •4

$\chi^1 = 4\cdot5$
$\chi^2 = 5\cdot8$
$4 = 7\cdot2$
$5 = 7\cdot9$
$7 = 8\cdot4$
$11 = 8\cdot9$
$14 = 9\cdot2$
$21 = 9\cdot7$
$29 = 9\cdot9$
$39 = 10\cdot6$
$45 = 11\cdot2$
$62 = 11\cdot6$
$99 = 12\cdot3$

•62   •39

•11

•99

•14   ⊙U ⊙UW   •21
•29

•45

U Orionis. A famous red Mira-type variable; magnitude 5.4 to 12.6; 372 days. Start from Zeta Tauri (page 231) and locate the pair of stars Chi[1] and Chi[2] Orionis; from these, identify the star 11 (magnitude 8.9). The second chart (inverted for telescopic use) shows the field round 11. Comparisons: Chi[1] (4.5), Chi[2] (5.8), 4 (7.2), 5 (7.9), 7 (8.4), II (8.9), 14 (9.2), 21 (9.7), 29 (9.9), 39 (10.6), 45 (11.2), 62 (11.6), 99 (12.3). Beware of UW, which is a Beta Lyrae eclipsing binary with a range of from 10.9 to 11.8 and a period of one day.

U Orionis is awkward inasmuch as its period is only a week longer than a year, and at the moment it reaches maximum during northern summer (about June/July) when it is too near the Sun to be seen. It reaches maximum about a week later each year.

I have left until last the baffling variable *R Coronae*, which has a range of from 5.8 to about 15. When at maximum – that is to say, for most of the time – it is on the fringe of naked-eye visibility, and is a binocular object; compare it with M (6.6) or 1 (7.2). If you look for it with binoculars and cannot find it, you may be sure that it has suffered one of its deep, unpredictable falls. If so, you will need a large telescope and a specialist set of charts to locate it.

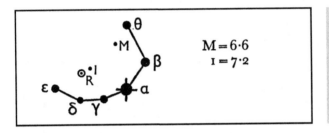

R CORONAE

Some Periodical Comets

No.	Name	Period (years)	Distance from Sun (astronomical units)		Next return
			Min.	Max.	
2	Encke	3.3	0.33	2.21	
26	Grigg–Skjellerap	5.1	0.99	2.96	
10	Tempel 2	5.5	1.48	3.30	
41	Tuttle–Giacobini–Kresak	5.5	1.07	3.10	
46	Wirtanen	5.5	1.07	3.10	
9	Tempel 1	5.5	1.50	3.12	
37	Forbes	6.1	1.57	3.42	
7	Pons–Winnecke	6.4	1.26	3.44	
6	D'Arrest	6.5	1.35	3.40	
67	Churyumov–Gerasimenko	6.6	1.30	4.0	
19	Borrelly	6.8	1.37	3.59	
15	Finlay	6.9	1.03	3.64	
54	de Vico–Swift	7.3	2.15	3.77	
4	Faye	7.3	1.65	3.78	
14	Wolf	8.2	2.41	4.07	
36	Whipple	8.5	3.09	4.17	
32	Comas Solà	8.8	1.85	4.27	
40	Väisälä 1	10.9	1.80	8.02	
8	Tuttle	13.5	1.00	5.67	
39	Oterma	19.5	5.47	7.24	
27	Crommelin	27.4	0.74	17.4	2011
55	Tempel–Tuttle	33.2	0.98	10.3	2031
38	Stephan–Oterma	37.7	1.57	20.9	2018
23	Brorsen–Metcalf	70.6	0.48	17.8	2059
12	Pons–Brooks	70.9	0.77	33.5	2024
1	Halley	76.0	0.59	35.3	2061
109	Swift–Tuttle	135	0.96	51.7	2126
35	Herschel–Rigollet	151	0.75	56.9	2090
153	Ikeya–Zhang	361	0.51	–50.8	2363

Appendix 28

Amateur Observatories

Some excellent modern telescopes are portable, but obviously a permanent observatory is much better; carry telescopes around, and it is only a question of time before something is dropped. A dome is ideal, but is not easy to make. The design which I would not personally favour is that in which the dome revolves; there is so much mass to be moved that jamming is inevitable. In a better design, only the top part of the dome revolves. The main trouble arises from the need for a circular rail.

Run-off sheds are much easier, and in general are perfectly satisfactory. My own is made in two sections. Each run back on rails that are connected in (angle-iron will suffice if need be). The shed itself is of wood, but hardboard is satisfactory enough. If the shed is made in one piece, one has to have a door at one end, and this, in my view, is not a good idea. If hinged, it will flap awkwardly; and to remove it entirely is not easy when one is working in the middle of the night and one's hands are cold. Moreover, any sort of door may tend to act as a powerful sail in high wind. The construction of a two-piece run-off shed is a sheer problem in carpentry, and the photograph should give adequate guidance.

Another method is to have an observatory in which the roof is run back on rails – the ends of the rails being supported, or by being fixed to the tops of poles concentrated into the ground. If this pattern is adopted, it is wise to make the roof as light as possible; plastic will suffice. The run-off roof idea is best suited for refractors, which have to be higher than reflectors and for which a run-off shed would need to be inconveniently tall. Remember, wind-force is a factor to be borne in mind.

Great care should be taken in the choice of a site for an observatory. According to the principle of Spode's Law ("If things *can* be awkward, they *are*"), trees and houses are always in the most inconvenient possible positions. If you can, select a site which is not only away from obstructions, but also well away from artificial lights, and from houses – that will give out warmth, so ruining definition. Above all, never put an observatory on top of a dwelling-house; flat roofs may look tempting but are to be avoided. A rooftop observatory has the worst of all possible worlds. It will experience the full force of the wind, and there will be so much warmth rising that no useful work will be possible.

Inevitably, the available sites will be far from ideal; and it is a question of making the best of things. For instance: if you are interested chiefly in the Moon and planets, select the site with the most favourable southern horizon. Inconvenient artificial lights can sometimes be screened. Even in my home in Selsey, within sound of the sea on the Sussex coast, I have had to put a screen in my garden to shield one awkward street-lamp. Reluctantly, I rejected the idea of using an air-gun to extinguish it permanently!

Appendix 29

Astronomical Societies

Anyone who takes more than a casual interest in astronomy will be wise to join a society. You will exchange ideas and observations with others – and you will make many new friends.

In Britain, the leading amateur society is the British Astronomical Association, which was founded in 1890 and has an observational record second to none. Its address is Burlington House, Piccadilly, London W1J 0DU (www.britastro.org). No qualifications are needed for entry; monthly meetings are held from October to June, and there are also many meetings held outside London. There is a bimonthly journal, and there are various sections each of which is in charge of an experienced Director. There is also the Society for Popular Astronomy and many local societies; a full list is published annually in the Yearbook of Astronomy (Macmillan and Co). Ireland has its national society; there are many in the United States. Elsewhere, there are many amateur members of the Royal Astronomical Society of Canada, the Royal Astronomical Society of New Zealand and the South African Astronomical Society. Neither are amateurs excluded from professional organizations, and indeed the list of Past Presidents of the Royal Astronomical Society includes the names of several amateurs.

Bibliography

So many books on astronomy are published each year that it is difficult to keep track of them, and all I propose to do here is list a few which have an observational bias and will not go quickly out of date.

ARNOLD, HJP. *Astrophotography*, George Phillip, 2002

ARNOLD, HJP, DOHERTY P and MOORE P. *Photographic Atlas of the Stars*, Institute of Physics, 2001

MOORE, P and LINTOTT, C. *Astronomy for GCSE*, Duckworth, 2002

MOORE, P. *Atlas of the Universe*, George Phillip, 2003

Various. *Small Astronomical, Observatories* (Vols 1 and 2), Springer Verlag, 2001 and 2002

Periodicals include The *Yearbook of Astronomy*, published annually by Macmillan. Monthly titles include *Sky and Telescope* (USA), *Astronomy* (USA), *Astronomy and Space* (Ireland) and *Astronomy* Now (UK).

Index